Sleek dugouts were beached in front of a Songhees summer fishing camp near Victoria, circa 1870. Dally photo, Royal British Columbia Muse

FOREWORD

Leslie Lincoln has a passion for the sea. I first came to know her when she came to my office in the Burke Museum in the hope of setting up a graduate degree program that could take full advantage of her live-aboard ketch *KHOYA*. It was instantly apparent that whatever direction her research took her it would be seaward and the course she eventually set, the study of Northwest Coast canoes, was inevitable.

The long hours in libraries and archives, digging through the journals of early explorers and the ethnographies of Northwest Coast people were balanced by journeys to the source—Salish craftsmen carving modern racing canoes or old-timers with tales of canoe travel from their own or their grandparents' experience. *KHOYA* took Leslie, whenever possible, to isolated villages where an old canoe might lie or where someone knew canoe lore. She searched out existing canoes in museums and parks and learned the mysteries of tables of offsets, stations and buttock lines in order to graphically record the forms of the canoes she found.

Leslie Lincoln's unbounded enthusiasm for her subject shows on every page. If any study ever was a labor of love, this one surely is!

Bill Holm, Curator Emeritus
Burke Museum

COVER PHOTO — McCurdy photographed the classic Coast Salish canoe off Port Townsend Bay, approximately 1906, Jefferson County Historical Society.

MAP I

Tribes of the Central & Southern

COAST SALISH AREA

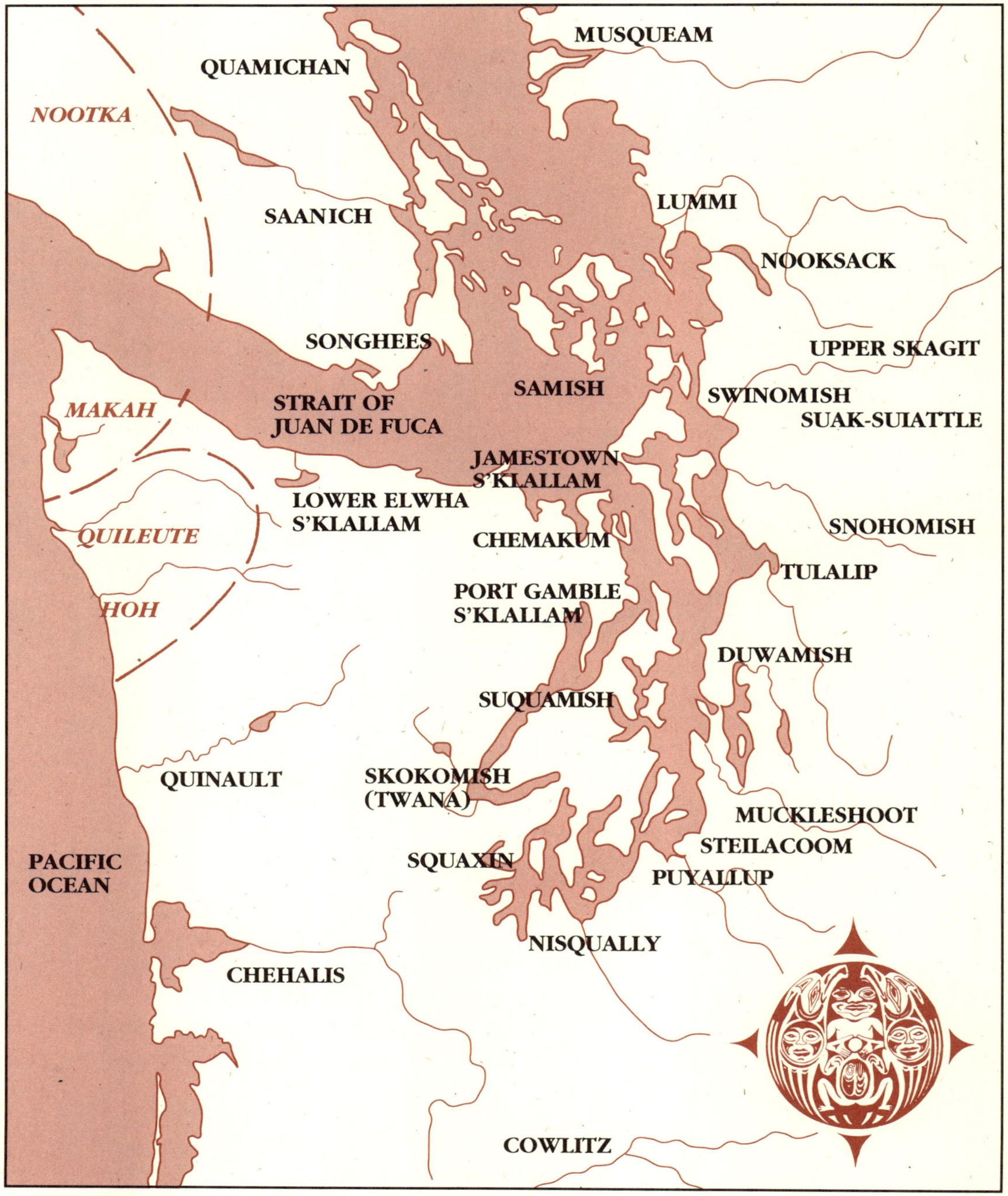

Quiet water and inland canoes of the Quamichan winter village near Cowichan Bay, B.C., 1866, Royal British Columbia Museum PN 1459.

CONTENTS

MAPS

AN INTRODUCTION TO THE COAST SALISH

Twelve years ago while anchored off southeast Vancouver Island, I heard an unforgettable song across the stillness of the evening water. This was my first encounter with a Coast Salish cedar dugout and crew. The steersman set the pace with his rhythmic call while the rest pulled their paddles in perfect synchrony with stirring speed and power.

The men's deep voices joined together the very moment the paddles switched sides. The canoe quickly passed out of sight but the memory has not faded. It has inspired a deep interest and respect for Coast Salish watercraft and culture.

Many Native people who live along the inland waters of what is now known as western Washington and southern British Columbia share the same linguistic roots and so are known collectively as the Coast Salish. In Washington this includes the Lummi, Nooksack, Upper Skagit, Sauk-Suiattle, Samish, Swinomish, Tulalip, Snohomish, Duwamish, Muckleshoot, Puyallup, Nisqually, Steilacoom, Squaxin, Skokomish (or Twana), Suquamish, Port Gamble S'Klallam, Jamestown S'Klallam, Lower Elwha S'Klallam, Chehalis, Cowlitz and Quinault peoples.

A few groups of canoe-using western Washington Natives do not share

David Current photo, 1988, Suquamish Tribal Archives WB 21B-37.

the Salishan language roots. They are the Makah, Quileute, Hoh and Chemacum. The Makah originate from the Nootkan language family centered on the west coast of Vancouver Island. The Quileute, Hoh and Chemacum people are grouped according to their ancient Chemakuan tongue. Their canoeing histories are included here because they have become a part of the neighboring Coast Salish dugout culture.

Most Coast Salish are inland salt water and river people. They continue to travel, trade, intermarry, gather and fish the areas now called Puget Sound, Admiralty Inlet, the Strait of Juan de Fuca, Georgia Strait and a small section of the Pacific Coast. Their graceful dugouts are this region's first traditional smallcraft.

Before a revival in 1989 few of the ancient-styled voyaging craft were still used, though sleek racing dugouts endured. Racing crews compete in a Salish circuit hosted in Indian country each summer.

The racing area extends into Puget Sound, though its center is northward in the Lummi and Central Salish reserves.

The Quileutes and Quinaults on the more isolated rain coast also continue to carve cedar dugouts. They use theirs equipped with outboards on the rivers for both fishing and racing.

In 1988 a Quinault fisherman, Phil Martin, told of motorized dugout racing: "When they get planing, it's exciting! It's a thrill—almost 30 miles an hour! The planing surface is 18 inches wide and 4 and a half feet long. The rest of the canoe's out of the water."

Canoe-borne, the Salish fishing people lived on nutritious resources from the inland sea, intertidal estuaries and rivers. They learned to catch, prepare, smoke and dry many types of fish, sea mammals, shellfish and waterfowl. This natural abundance coupled with their expertise, created a rich fishing and gathering society. They frequently produced commodities beyond immediate need. Though hunger lurked behind a poor fish run, the usual bountiful catch of returning salmon gave the Coast Salish peoples the opportunity to refine their social and ceremonial ties, their song, dance, spirit quests, weaving and carving. The canoe with its fair profile, split bow, delicately flared gunwale edge and upturned stern is an elegant symbol of their cultural achievements.

The hardy dugouts made this marine lifestyle possible — indeed, successful. Their carrying capacity allowed for seasonal migration from central winter settlements to summer fishing camps. Tons of dried and smoked salmon, halibut, cod, smelt, shellfish and seaweed were transported back to the winter houses located near the shore. Intertidal beach clearings and canoe skids marked village or camp landing sites.

The waterways coursed with fleets of canoes. Whole villages in large ceremonial craft voyaged to social gatherings, dances and potlatch feasts. This extensive water travel created a lasting fabric of marriages and related families. Offshore sea mammal hunting, slave raids and retaliatory warfare demanded further performance from the seaworthy craft.

HISTORICAL NORTHWEST COAST DUGOUTS

Canoes surrounded Captain Meares' longboat off the Makah coast, 1788. Special Collections, University of Washington Libraries NA 3959.

According to geological evidence, migration across "Beringia," an exposed land bridge connecting Asia to the New World was probable during the last major glaciation 10–15 thousand years ago. The first human inhabitants likely crossed that land bridge and then continued their trek southward in ice-free valleys.

Alternatively, archaeologists De Laguna and Fladmark suggest early peoples may have used skin-covered or bark-built boats to explore river valleys or coastal regions. The 9,000 year old remains of sea mammal fisheries and occupation on the isolated Queen Charlotte Islands are convincing testimony for an early advance of watercraft.

Through these waves of migration or through supernatural intervention as Native stories suggest, aboriginals arrived on the northwest coast. Once here, they encountered dense forests and particularly the great western red cedar, *Thuja plicata.* It is light in weight, easy to work, yet highly durable. Its bark can be shredded and worked into clothing,

waterproof hats, baskets and plaited mats. Its branches and roots can be worked into strong ropes. The wood splits naturally into house planks, bends into watertight boxes and is hewn into the graceful canoes. Within the cellular structure of red cedar is an organic toxic oil, called thujaplicin which acts as a preservative to resist rot in mature trees. This oil contributes to the tree's pleasant aroma. Cedar is the "Great Life Giver," the Indian's sacred tree.

For more than 2,000 years, perhaps as many as 8,000, the Coast Salish peoples carved a vital life on these shores—rich with sophisticated fishing and gathering technology, voyaging skill, several Coast Salish languages, many distinct dialects, as well as unique religious, ceremonial and art traditions.

These were the legendary times—epochs of supernatural power and creation. Elders still share the legend of the Great Flood. Similar to the biblical story of Noah's Ark, chosen ancestors were saved from a rising flood in a great canoe. A tremendous cedar rope lashed them to a mountain peak. In the tempest the rope broke and the canoe drifted freely. After the waters receded the survivors settled near where their canoe landed. A Lummi story elaborates:

> *In a giant crack halfway up the slope of Mount Baker is the outline of the giant canoe. It has been there ever since the great flood (Clark 1953).*

These extraordinary times are considered "prehistory" occurring before the written word. Humanity's ingenuity was captured in stories, inscribed in artistic creations and coded in material objects. Their canoe designs record this legacy—generations of seafarers who refined and perfected the vital art of canoe making.

Traditionally the master carver hewed a dugout from a single log—generally red cedar although carvers also made use of yellow cedar, spruce or cottonwood in upper river areas where these were more available. In length the canoes varied from 8 to 72 feet though most measured between 15 and 35 feet. They knew how to steam spread the hull of the canoe log to increase its beam and stability.

Along the entire forested Pacific Northwest Coast from Yakutat Bay in Alaska to northernmost California, a variety of related types emerged. The different hull shapes served a range of needs, from open ocean to inland and river waterways. Over many, many moons these styles were perfected to handle a variety of wind, wave, sea and current conditions.

A canoe model was unearthed at Ozette, dated to approximately five hundred years ago. Even at that time the Nootkan style model shows characteristic features: a projecting animal-form bow, a carved projection or "Adam's apple" along the neck and a vertical stern post. This long-term continuity reveals conservative stability in design.

The great war canoes disappeared with the arrival of American and British armed vessels in the early 1800s.

Coast Salish and "Munka" war canoes drawn by Paul Kane, 1841. Thomas Burke Washington State Memorial Museum.

In the 1700s during the era of encounter, Native peoples met Russian, Spanish, British, French and American naval envoys sailing in their ships. Makah storyteller Helen Peterson recalls one such encounter when the village people came upon a large "floating house" just drifting in. The Makah still refer to whites as those who "drift in."

Ship's logs describe details of these encounters and shipboard artists illustrated the early canoe forms. John Webber, illustrator for Captain Cook, documented the Salish and Nootkan types in 1778. The British Museum houses two miniatures collected by the Vancouver expedition in 1792: a Nootkan style model and the war canoe model known in the literature as a Head canoe.

Many large war canoes surrounded Captain Gray's vessel, the *Columbia,* in 1792. Fleets of Head canoes confronted naval explorers who reported encounters with them from Lituya Bay southward to Gray's Harbor.

Once plentiful, this Head canoe style war craft has long been extinct, though shipboard officers chronicled exploits with the craft in 1770–1830, the contact period. A replica is being built by Marvin Oliver and Steve Brown in Seattle at the United Indians of All Tribes Foundation. When this one is launched and tested it will re-awaken a lost legacy.

The Head canoe is named for its unusual bow, an elongated fin carved into the front or "head" of the vessel. The strikingly rectangular bow may have served as a shield in warfare, though its purpose is a mystery. The stern, also a thin, fin-like blade, projects at an angle up and outward over the water. Because canoes were pulled up on the beach stern first, ready for launching, this angled stern fin would not have been damaged by such landings.

Another extinct war canoe is referred to as the "Munka." In hull form it was related to the Nootkan style dugout. Based on paintings from early explorers'

Fleets of warriors in Head canoes challenged the COLUMBIA in 1792. George Davidson drawing, Oregon Historical Society.

expeditions, the Munka plied the waters from Nootka Sound to as far south as the Columbia River.

Models have been collected, though no full-size vessels remain. Both the Head and Munka type disappeared in the early nineteenth century probably due to confrontation with the cannons of American and British vessels.

Yet a dugout style remains which closely resembles the form of the ancient Head canoe. This is the spruce canoe and it continues to be built for special cultural projects in northern areas such as Glacier Bay.

Both the ancient Head canoe and its modern counterpart, the spruce canoe, have distinctive fin ends though the latter's extremities are shorter. The spruce canoe, popularly known as the Sitka canoe at the turn of the century, was also carved from cottonwood. In 1874 a spruce canoe model was collected in Port Mulgrave and photographs show the cottonwood version along the Chilkat River.

The craft was carved where cottonwood and spruce were more available than cedar. A river variant with fin ends may yet be found in selected reserves of the coastal inlets and river valleys of British Columbia.

Fishermen along Alaska's Chilkat River dip-netted for the prized eulachan. Spruce style canoes are closely related in form to the extinct Head canoes. American Museum of Natural History #13989.

NORTHERN STYLE CANOE

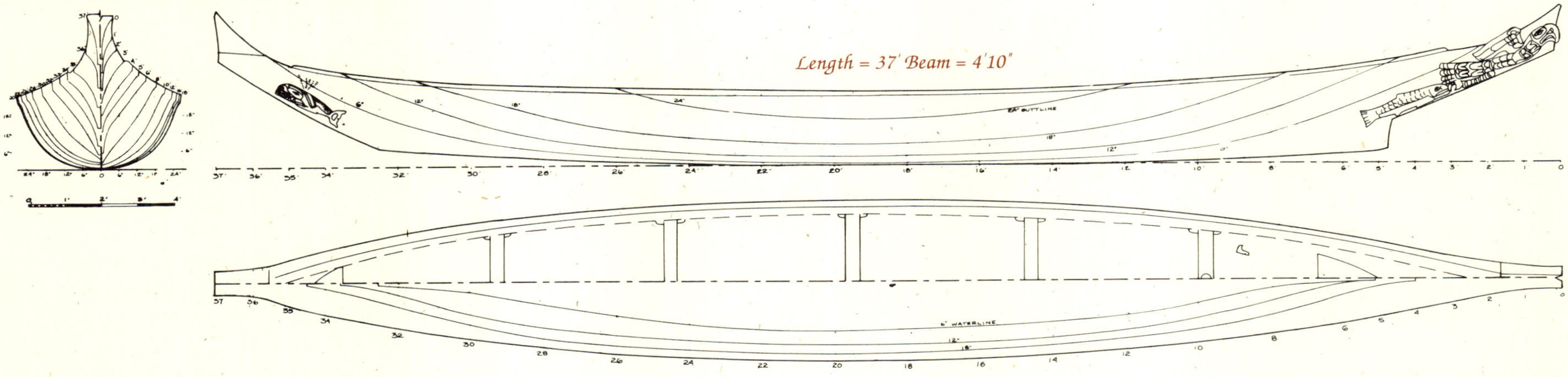

This Northern style (Kwakiutl) dugout was collected by Mrs. Burke in 1896 from Chief Jonathon Whonnock of Alert Bay, B.C. It is displayed at the Thomas Burke Washington State Memorial Museum, Seattle. Lines taken by Bill Holm and drawn by Leslie Lincoln, 1989.

COAST SALISH STYLE CANOE

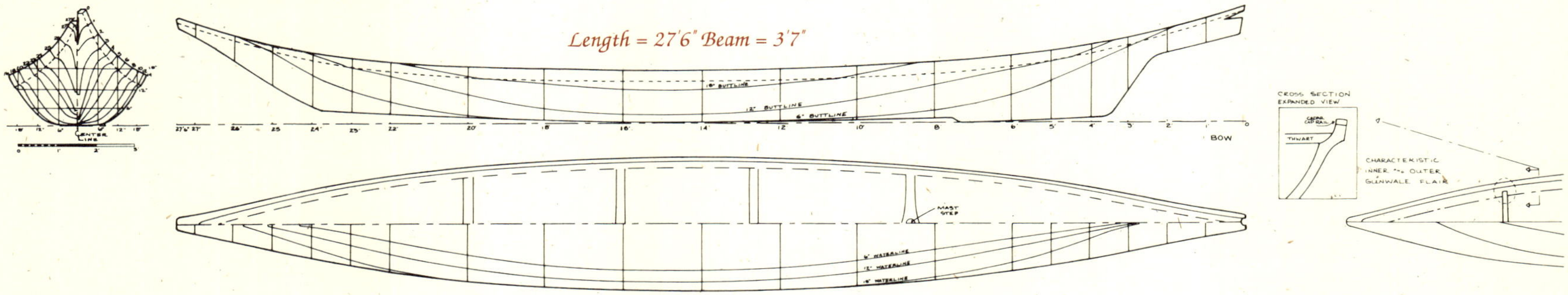

This classic Coast Salish canoe is housed in storage at the Vancouver Centennial Museum, Vancouver, B.C. Lines taken by Duane Pasco and Leslie Lincoln. Drawn by Lincoln, 1988.

Coast Salish style canoes are an ancient form of watercraft evolved for use in the inland seas. The style is clearly portrayed in early artists' renderings, photographs and ethnographic descriptions. This sleek craft has a low freeboard and gently sweeping sheer—or in non-nautical jargon, it has a profile with gracefully uplifting front and back ends. The prow of the Coast Salish style canoe has a notch in the very end. In 1920 T.T. Waterman and Geraldine Coffin wrote a fine book called *Types of Canoes on Puget Sound*. Their informants said the split bow resembles "an open mouth." Its stern is said to "turn up like a duck's tail!"

The Coast Salish split bow form has dual functions. The upper edge flairs, forming a buoyant horizontal surface which resists boarding waves. The lower vertical portion of the split bow cuts downward, angling toward the cutwater or forward fin. This cutwater slices into waves for a fine entry and is usually less than one inch wide. On several existing Salish canoes remaining in museums, the bow cutwater extends downward half an inch deeper than the bottom of the hull. Waterman suggested that this subtle keel-like extension enabled Native hunters to travel silently and so surprise sleeping prey. It also increases the hull's tracking ability.

Jim Dick of Squamish, B.C., navigated this Coast Salish freighting canoe in 1928. H.I. Smith photo, Canadian Museum of Civilization #71529.

There are two basic variations of this Coast Salish type—one for carrying cargo and one for hunting. The heavier, freighting version was recorded in Lushootseed, the Puget Sound Coast Salish language as "Sti'whahl." It reached lengths up to 40 feet.

The sleeker Coast Salish style variation was known as the "Sda'kwihl." The Sda'kwihl was intended for porpoise, seal and duck hunting. From linguistic evidence, Salish anthropologist Wayne Suttles explains that the term "Sda'kwihl" clearly identifies this type as the most common canoe in the area. The widespread usage of this Native term implies great antiquity of the canoe style.

A groove runs fore and aft on the inside of the hull's topmost edge, the gunwale. This groove complements a delicate out-flaring edge along the outside of the hull. This looks like an added plank or washstrake though it is integral—adding buoyancy to the design.

Holm suggests the Salish canoe form inspired the development of the Northern hull shape which was first seen and documented in the nineteenth century. The distinctive Coast Salish style cutwater, gunwale rim and uprising stern are all paralleled in the often larger Northern canoe design.

A sweeping sheer, uprising bow and stern as well as high sides, or freeboard, distinguish the Northern style cedar dugouts. The Tlingit, Haida, Tsimshian, Bella Bella, Haisla and Kwakiutl each crafted their own versions of this style of seaworthy craft to meet the demands of their exposed waters.

On special ceremonial canoes carved images adorned the prow and stern. The owners had their clan figures— raven, eagle, killer whale, bear

Sixty foot Northern style canoes could carry as much as six to eight tons of freight.

or frog — painted on the bow and stern sections in the striking Northwest Coast formline designs.

Emissaries journeyed to invite neighboring peoples to gather at potlatch give-aways. When arriving from a distant village it continues to be necessary for a crew to announce themselves and their peaceful intentions. On a plank across the bow dancers in abalone frontlets and button blankets or Chilkat robes would sing their hereditary songs before an assembled crowd. Lead dancers might shake sacred eagle down as a sign of peace.

The Haida on the Queen Charlotte Islands are expert canoe makers. Great stands of large cedars enable them to make outstanding vessels as much as 60 feet in length with two masts and sails. Once the rugged mariners traded and delivered them to neighboring peoples. *In the Wake of the War Canoe,* by William Collison, tells the remarkable story of the first Haida missionary's canoeing perils and adventures at the turn of the twentieth century.

Northern warriors were infamous for their raiding exploits. They also were ready voyagers and traveled extensively

This northernmost type, the Yakutat sea otter hunting canoe, plied the icy waters of Yakutat Bay. Harriman photo, 1899, Special Collections, University of Washington Libraries NA 2098.

on peaceful trading missions and to work the hop fields near Seattle. Bentwood boxes with tight fitting lids and waterproof bags of halibut skins or sealskins carried their personal belongings.

The hull of the Northern canoe is almost rounded at its bottom amidships. Flaring sides offer increased stability particularly when the canoe is loaded. Small pieces are scarfed onto both the bow and stern extremities to add further height to these commanding vessels.

In the northernmost waters of the Northwest Coast the Yakutat seal or sea otter hunting canoe was found. This style is distinguished by an unusually prominent bow cutwater. There is also a fin on the stern. These small 16–18 foot hunting canoes were hewn from spruce or yellow cedar if a suitable drift log was available. De Laguna recorded that the fin on the bow held the canoe steady in rough seas or swift currents and turned aside patches of ice.

Northern style canoes brought Tlingits together for a July 4th gathering in Sitka, Alaska, 1904. Thomas Burke Memorial Washington State Museum 2.5 L133.

The fine, upturned shape of these Yukon Valley birch bark canoes may have influenced the design of Northern style dugouts which emerged in the 19th century. Photo circa 1898, Yukon Archives, Alaska Historical Library collection #4232.

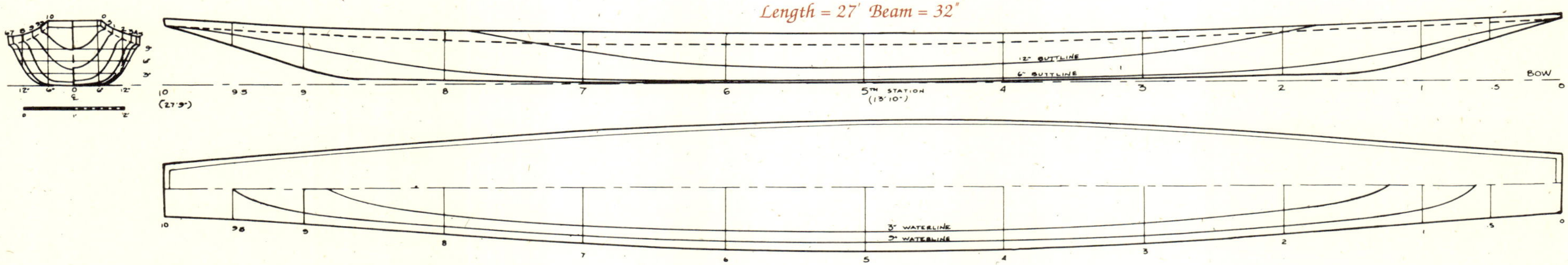

COAST SALISH SHOVEL-NOSE CANOE

ABOVE — This cedar dugout was poled along the Stillaguamish River and now can be found at the National Forest Service, Darrington. Lines taken and drawn by Leslie Lincoln, 1988.

RIGHT — Shovel-nose canoes were ideally suited for gliding over swift cross currents. E.S. Curtis photo, circa 1912. Special Collections, University of Washington Libraries NA 306.

Shovel-nose dugouts were once essential to the river tribes around Puget Sound including the Nooksack, Upper Skagit, Sauk-Suiattle, Snohomish, Stillaguamish, Puyallup, Nisqually, Chehalis, Muckleshoot and freshwater Duwamish.

They generally traveled up or downstream from their villages in the "Tl'ai," that is, the shovel-nose dugout. They frequented river mouths, intertidal wetlands and inland tributaries. The shallow river craft ranges from 10 to 40 feet long.

The symmetrical scooped bow and stern ends for which it is named do not catch the rapid currents of swift moving rivers, yet are sufficiently buoyant to support a fisherman or woman with spear or net.

Many interior Salish peoples who had only the river canoes preferred the safety of an inland portage to crossing exposed saltwater regions. Typically these shallow vessels were poled to headwaters where goods and canoes were portaged to another nearby river or creek, following established interior routes.

In 1920 Waterman and Coffin suggested that the first style of dugout used on the Northwest Coast was the "shovel-nose" type and that it preceded the development of all later watercraft. Yet the Tlingit made use of a different style, the sharp bow cottonwood river canoe. In a material culture survey Drucker found no northern Northwest Coast groups had shovel-nose canoes, which implies that the inland shovel-nose was in fact not widespread, nor was it likely the universal forerunner.

In the north only one group, the Bella Coola, did use a variant of the shovel-nose style — the "Spoon" type. This craft with upturned bow and stern is

The Yurok canoe, the southernmost shovel-nose, was hewn of redwood with an added-on vertical transom. Museum of the American Indian Heye Foundation, 1920.

named for its resemblance to goat horn spoons. It is noteworthy that the Bella Coola are members of the Salishan language family. Their canoe style reflects these Salishan roots. The Kwakiutl located between these two Salish peoples used a pointed or sharp bow river dugout. Some Kwagiulth people in the remoter areas still use the sharp bow dugout equipped with outboard power.

Waterman emphasized a similarity between the Puget Sound Salish shovel-nose canoes and the southern Yurok canoe: both are square-ended and highly refined but there are differences. Dugouts of the southern Oregon and northern California peoples were made of durable redwood where this became more available than cedar. The Yurok's undercut, inward rolling gunwales provide a handhold for portages and beaching. Its bow and stern are raised by adding pointed vertical transom pieces.

To the southern limits of the coastal redwood forests the Yurok style shovel-nose canoes became less common. Log rafts and large basket-style coracles served the needs of water transport along with some roughly hewn dugouts.

Beyond the southern limit of dugout territory the Chumash peoples of central California produced plank canoes, the only indigenous wooden planked watercraft of North America. But that is another story.

Bella Coola river dugouts are known as "Spoon canoes" for their resemblance to goat horn spoons. Royal British Columbia Museum PN 4601B.

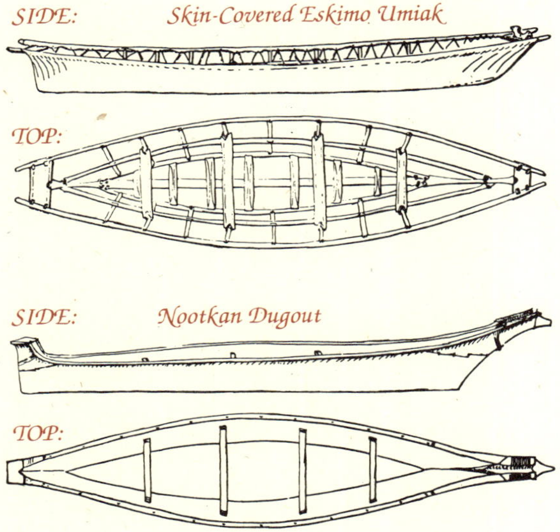

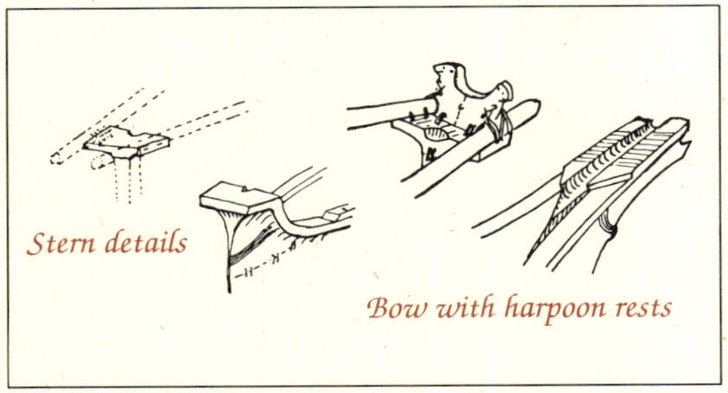

Through a cultural and stylistic comparison, Duff proposed that the Nootkan dugout design evolved from the Eskimo skin-covered Umiak. Both open ocean hunting craft share similar structural features. Royal British Columbia Museum drawing by D'Oyly Rochfort, originally published in 1964.

The Nootkan canoe, also known as the Westcoast, whaling, traveling, deerhead or Chinook type, is famous for its whale hunting use by the Makah and Nootkan (or Westcoast or Nuu-cha-nulth) people. This type was first built along the outer coast of Vancouver Island. There the Nitinaht and Clayoquot were famous for their large cedar stands and skillful canoe carving.

The Salish Lushootseed term for the ocean-going dugout can be analyzed in Nitinaht, a Nootkan language. The term means "whaling canoe," though it is not decipherable in Lushootseed. This is linguistic evidence for the canoe's Nootkan origin.

The ocean-going dugout has an almost flat bottom with flaring sides, which results in stability and a good load carrying capacity. These able sea boats were so well-designed for freighting, fishing and sea mammal hunting that they were exported and became the dominant type of watercraft until the 1920s for the Salish area and inland along the Columbia River.

The ancestry of the Nootkan canoe form is further reinforced by its wide distribution, traded down the coast as far south as the Columbia River. Early explorers on this river called the type "Chinook" for the people there who used them.

In the 1800s Native sea mammal hunters worked on sealing schooners, taking their Nootkan style sealing canoes aboard with them. The projecting bows of these working dugouts were cut off so they could be stacked like dories aboard the larger mother ships. This extended the range of Nootkan canoes to the Aleutian Islands and Bering Sea.

Wilson Duff suggested the Eskimo skin-covered Umiak could be the direct ancestor for the design of the Nootkan canoe. There are startling similarities between the Eskimo and Nootkan whaling practices, preparatory rituals and the hull features of the two types of ocean-going craft.

Duff noted that details of the Nootkan canoe may be vestiges of once structural components for the skin-covered Umiak. Perhaps these structural forms were integrated into the Nootkan's unique stern and overall hull design.

The high projecting bow is a separate piece scarfed into the hull. Its shape is like a poised head with ears, snout and an Adam's apple or heart. Between the forward extremity or "ears" is a notch for harpoon or mast. The aft end gradually narrows in breadth to less than an inch wide on an almost vertical stern post also scarfed into the hull. Eugene Arima has written and illustrated a fine study of the entire process of carving and steaming a Nootkan craft in Port Renfrew in 1975.

Numerous Nootkan style canoes gathered along Seattle's waterfront on Ballast Island. This was the most common type found during the nineteenth century. J.P. Soule photo, circa 1888–1895, Special Collections, University of Washington Libraries NA 680.

COAST SALISH RACING CANOES

Songhees Indians escorted Lord Willingden on a Victoria Day parade, Royal British Columbia Museum PN 16760.

Racing continues as a vital tradition for many Salish people. In Washington the Lummi, Nooksack, Upper Skagit, Skokomish, Tulalip, Suquamish and Makah host and participate in annual circuits. In British Columbia the East and West Saanich, Capilano, Cowichan, Upper Stalo and Musqueam race. As many as 22 teams in the 50 foot eleven-person racing canoes vie for fun, prestige and prize money during weekend events.

Smaller dugouts, called singles, doubles and six-person 42 foot canoes compete. Women and teenagers race as well as men. The circuit lasts from May through August with races hosted by different Indian bands or reservations. Families gather on the beaches to lend support and enjoy the summer. Some sell crafted items and beadwork. Couples meet at evening dances. Salmon barbecues are shared and the gambling bone games go on into the night. The races bring separated tribal communities together. In the 1920s teams paddled their canoes to get to hosting villages. Today canoes are loaded atop trucks, vans and trailers for these get-togethers.

Crews have challenged each other as long as they have been afloat. Fleets assembled for winter dances, potlatch gatherings, abundant fish runs, at portage sites and during warfare. Feats of seamanship and quick passages have been cause for fame, prestige, rivalry and for survival.

Able crews were vital to the security of the village. Safe travel, trading and successful fishing trips, during stormy conditions or fog, required keen knowledge of "uncharted" waters, currents and tides.

In the nineteenth century, this sport celebrated those who regularly fished, traded and traveled in canoes. They earned distinction and renown in the finely crafted dugouts.

"Like the potlatching, it was competition with the canoes," Bill Blaney of Campbell River recalled. "We built our canoe for a visit from King George and Queen Elizabeth who watched the races from Stanley Park in Vancouver, B.C."

Missionary Myron Eels described his experience as canoe passenger before the turn of the century:

> *About eight o'clock we started and reached Port Townsend in about an hour. Here they spent nearly two hours in purchasing things to present to the principal men at the potlatch, and the day being pleasant we went on, having a race in which nearly all the canoes took part. As there was little wind it was a trial of strength and endurance and was engaged in for mere sport. It was kept up for two or three miles, until one canoe had passed the rest and the losers were satisfied that it was useless to contest further.*

Fifty foot canoes at the start of a 1930's race on Penn Cove, Island County Historical Society.

Since the virtual take over of other modes of transportation, competition between strong crews has taken on new meaning:

> *In the past, the canoes kept the people strong in the face of northern aggression. Today, they keep the people strong in preserving their sense of Indian identification in an increasingly non-Indian world (Hamilton 1980).*

The racing canoe is related in hull form to the Nootkan style dugout. The sleek 50 foot canoes show the distinctive stern and animated prow. Some used by the buckskins, or teenagers, are so narrow they are just wide enough to fit the crew's hips. There is little freeboard on these stiletto craft.

The racing design is shaped for speed. Some scholars say changes in hull form were influenced by collegiate rowing shells brought to the Northwest in the early 1900s.

Lummi racers explain the 50 foot hull attained its radical length and narrow

SALISH RACING CANOE

Length = 50'8" Beam = 2'6"

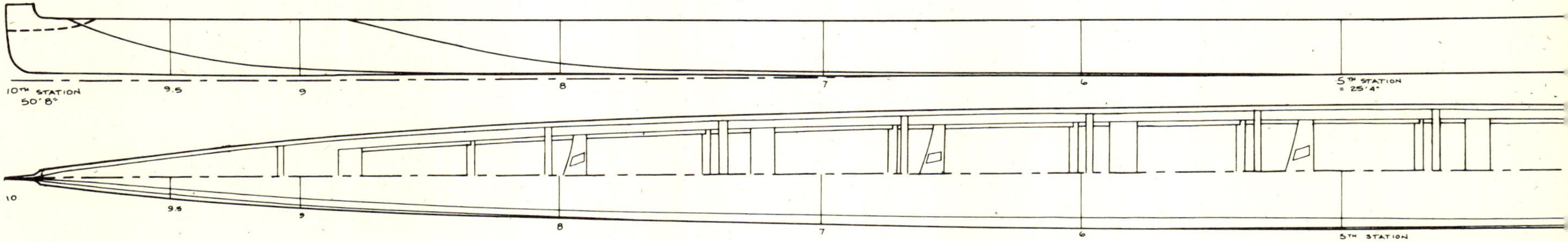

Charlie and Dick Edwards carved TELEGRAPH, a champion eleven-person racing canoe, in 1906 on the Swinomish Reservation. Lines taken and drawn by Leslie Lincoln, 198

NOOTKAN STYLE CANOE

Length = 25'6" Beam = 4'

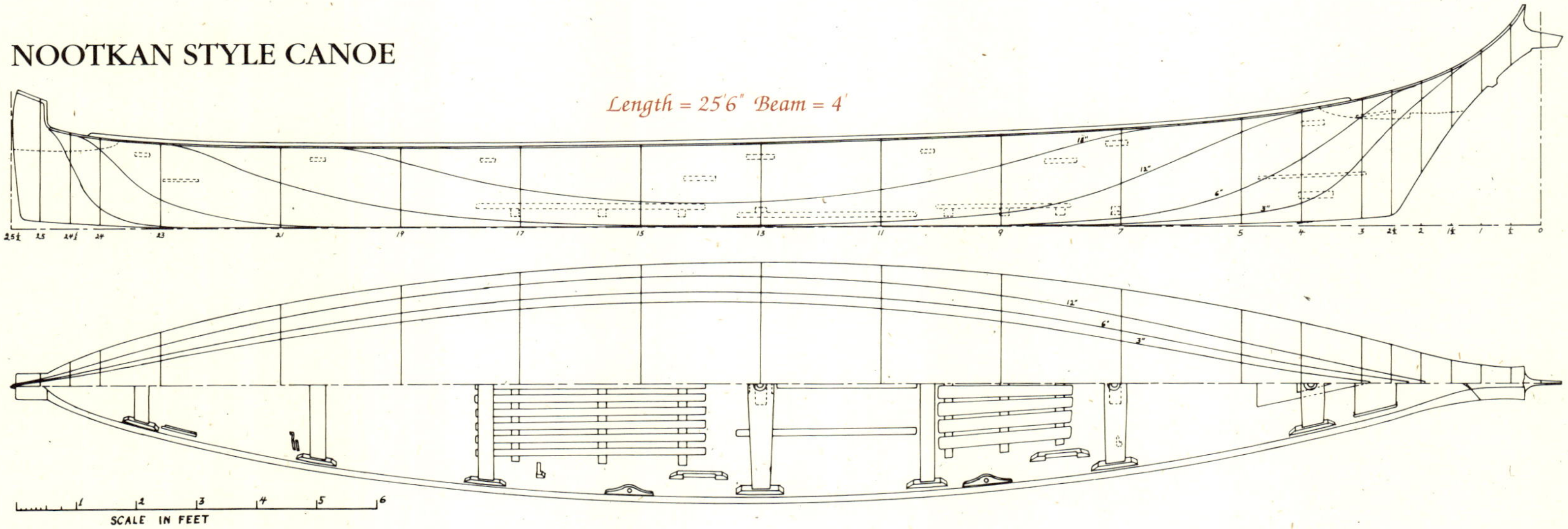

This Nootkan style sealing canoe was built by Conrad Williams, a Quileute at La Push, about 1928. The canoe was used for seal-hunting before being collected by Orre Nobles in 1930. Lines taken and drawn by Bill Durham, 1965.

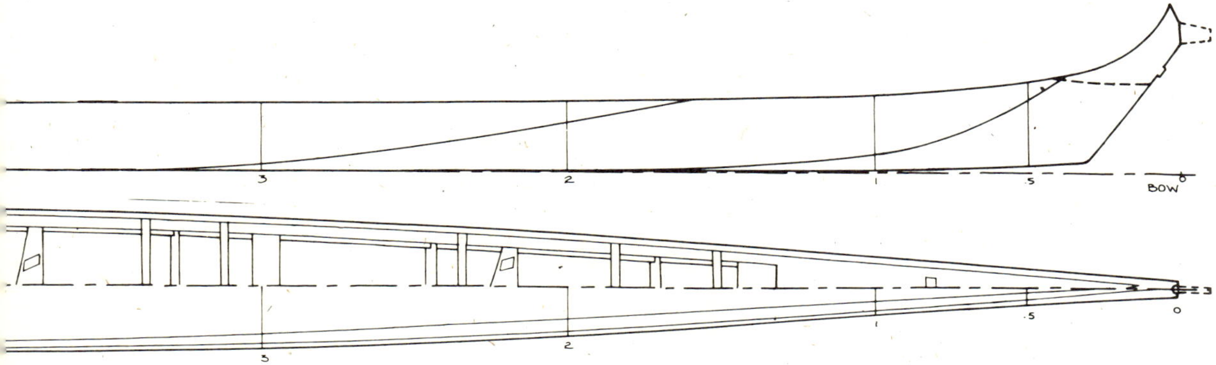

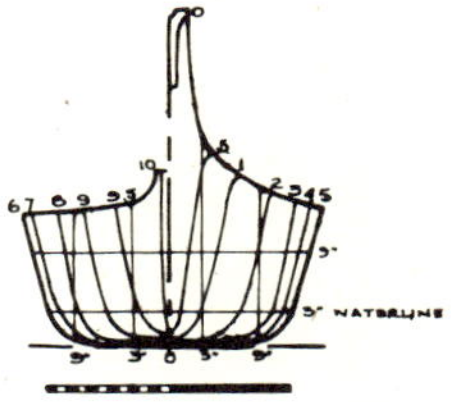

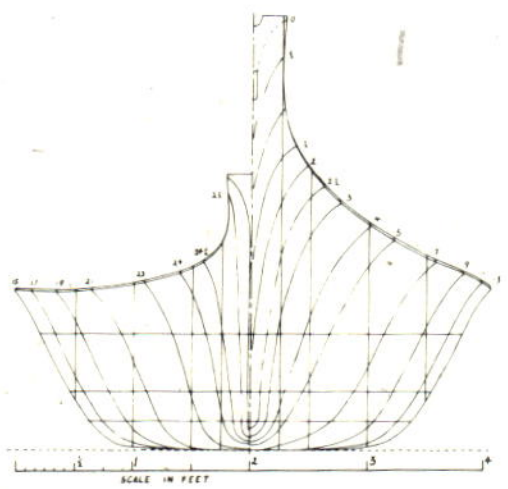

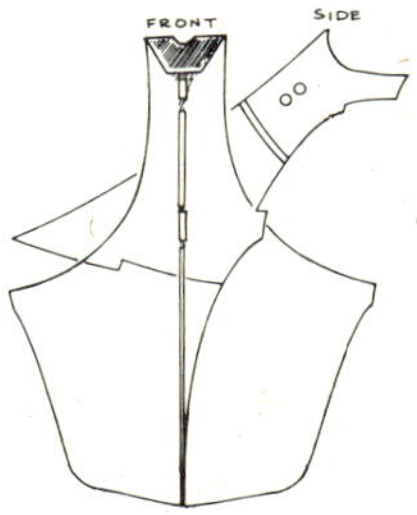

beam from the need to compete with northern raiders who came to steal their women and children in large sea-going craft. Salish carvers made their dugouts longer and faster to better intercept raiding vessels. Hence the popular term for them: "War Canoes."

The *Telegraph* was a prototype for the successful 11-person racing canoes built by the Edwards brothers. She was rushed to completion in about three weeks to challenge her rival *Valdez* in the big 4th of July races at Anacortes.

Despite her crew's best efforts the *Telegraph* proved too new and too heavy in the water. Her hull was thinned and by 1910 her crew handily defeated the *Valdez* on a five-mile course. The *Telegraph* remained champion for nearly 20 years until about 1930.

Finally the veteran wore so thin it was feared she would crack through on the bottom. So a sister ship, the *Question Mark,* was built to replace her, also performing well. Recently restored, the *Telegraph* now rests under a shelter, displayed by Coupeville's Island County Historical Museum.

A new Lummi racing canoe carries the name *Telegraph II*—bringing forward a tradition of success. The Lummi now challenge *Geronimo* from the West Saanich Tsarlip Band. *Geronimo* has won the 11-person race for the last seven years, holding title to the Coast Salish championship. A speed of 11 knots can be made by top crews.

Competing with as many as 20 canoes, TELEGRAPH and her crew consistently won the championship. She now rests on display in Coupeville. Near there, by Chief Sneaklum's place on Penn Cove, the popular races were held. Thomas Burke Memorial Washington State Museum #L-4258, July 4, 1920.

Before racing, ritual preparations for both the canoe and crew ensure confidence and success.

A buckskin, or teenager, competes in this single, a contemporary racing dugout. David Current photo, 1988, Suquamish Tribal archives WB2B #36.

A few Salish carvers continue to hew the long racing canoes. Roy Edwards of the Gulf Islands received his instruction from elders with hand tools and hand measurements. He is recognized as a master and has built more than 20 of the large craft for different groups, including the Lummi, Upper Skagit and Suquamish. Most carvers learn the techniques by building the smaller one- and two-person cedar dugouts.

Many racing canoes are shaped though not finished in the first year. The hull may be left as much as 3 inches thick as any slight warping or twisting is noticeable in a fifty foot craft. The following year a layer of thickness is adzed off, lightening the shell's weight. Over the years the canoe will continue to be thinned until ultimately she will need frames and patches for strength and structural integrity. The hulls are surfaced to a polished finish. Soapy water is sometimes rubbed onto the underwater section just before a race.

Debating the thickness of the hull, George Swanaset, a Nooksack racer, explained that with a heavier hull it's harder to get going at first but once underway she will go faster, carrying her momentum. So there are some advantages to having a heavier hull.

The hull is not the only concern. The crew and their training are most important. Native people who paddle call themselves "pullers." The term "pullers" may well have its roots in Straits Salish. According to linguist Thom Hess the verb "to paddle" in the Saanich dialect means, "to pull" and refers to the mariner's physical movement — the pulling motion.

Rick Edwards, a trainer for Lummi's women's crew, has pulled and trained others for a score of years. He comes from a strong heritage of canoe makers. His father, grandfather, great-grandfather and no doubt those before, also pulled in cedar watercraft. Edwards teaches racing at the Northwest Indian College. He shared some tactics.

Training begins in January with road-running, stamina-building and attention to healthful diets. No drugs or alcohol are allowed for the canoe crews. Fund raising projects are planned to finance weekend trips to hosting competitors throughout lower British Columbia and northwest Washington.

Training for the main men's team is particularly challenging. Seven years of

Women race in the Coast Salish championships. David Current photo, Chief Seattle Days, 1988. Suquamish Tribal Archives WB27A #17.

practice and endurance building are required to earn a seat with the champion crew.

The crew for the main men's competition is selected with care. Six of the eleven pullers are chosen from the best trained men and race both days. Five of the next-best race on Saturday and the next-five race on Sunday—Trophy Day. This gives those of the greatest stamina the privilege of racing both days, yet new strength of the other men is also gained.

Prize money is put up for the winning team. For example, in 1989, over $6,000 was put up for the Chief Seattle Day canoe races in Suquamish.

Veteran Lummi racer, Willie Jones, explained his philosophy of canoe racing:

How do you want to feel
And look when you get there?
I think you should be proud,
And look in time as a team...
I am really proud
When I step in that canoe...
The expectations and standards we set
For ourselves are excellence.
To me—that's Indian,
To me—that's human,
Being well.
I think them are the goals
We put for ourselves,
To be well and to be together.
Even if it's a small scale,
It's a model we can look at
And our community
Can take pride in that.
To me—that's the real winner,
Not coming first,
But them two things in combination—
HEALTHY and TOGETHER.

MAP II

NATIVE NORTHWEST COAST CANOE STYLES

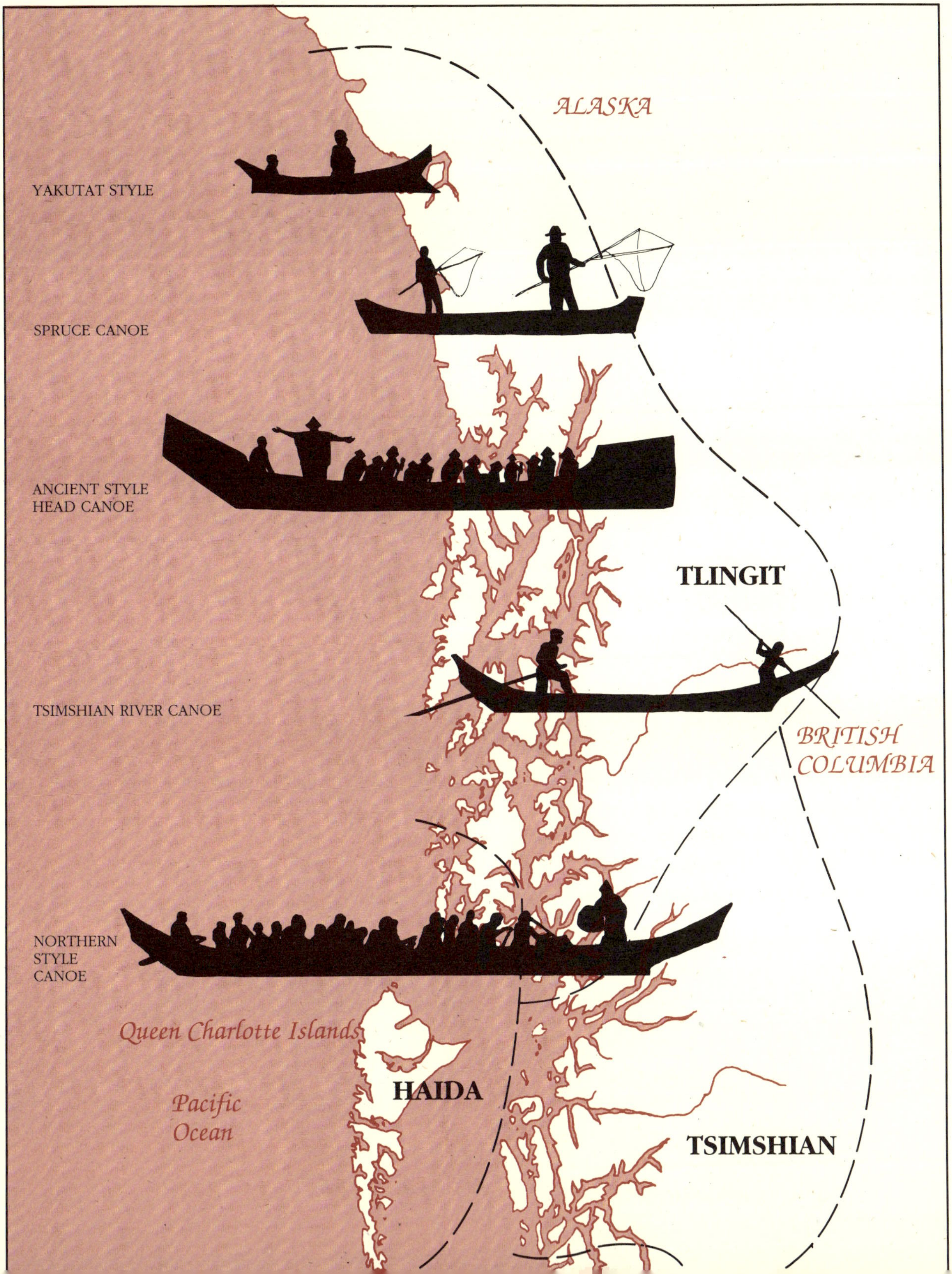

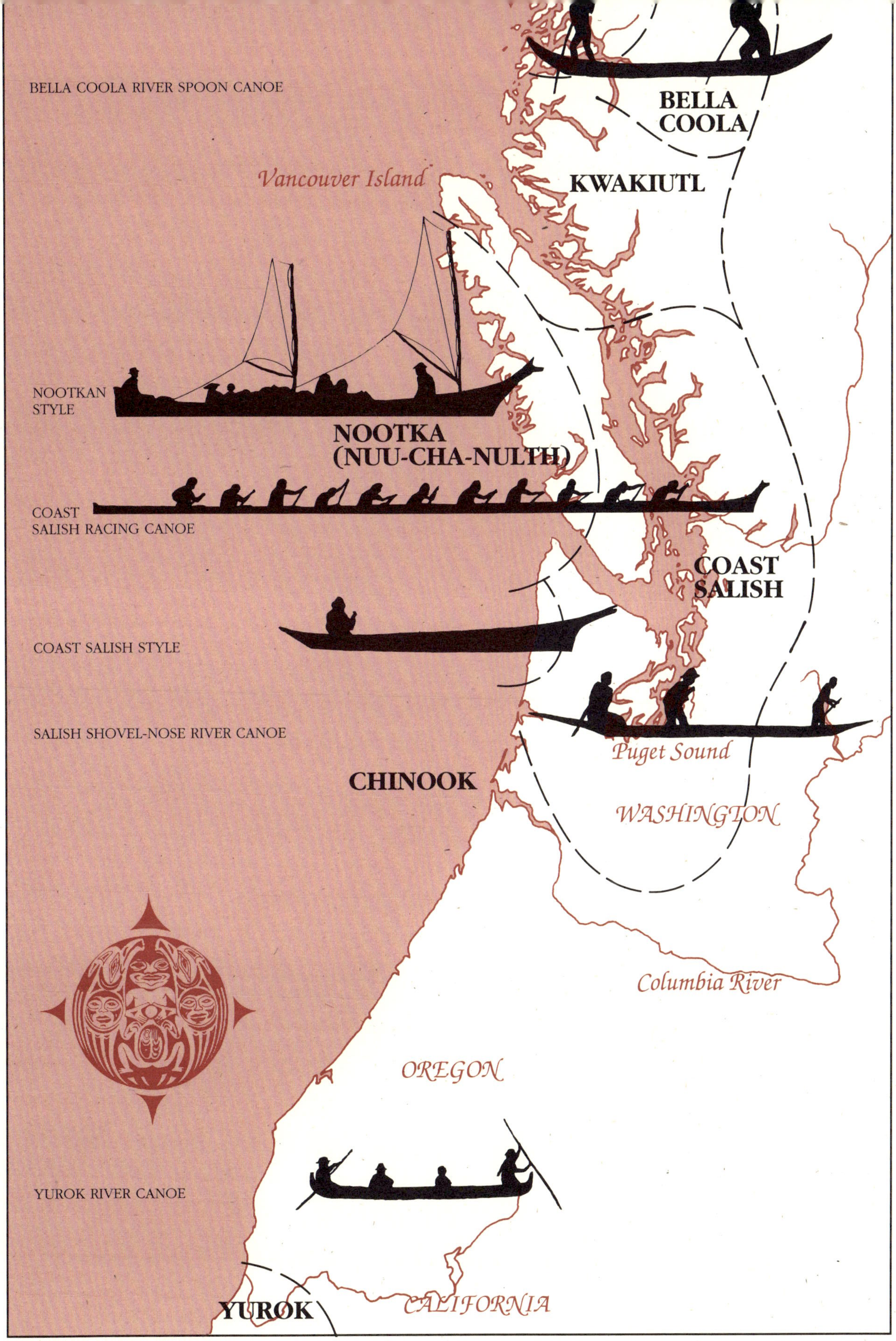

BELLA COOLA RIVER SPOON CANOE
BELLA COOLA
KWAKIUTL
Vancouver Island
NOOTKAN STYLE
NOOTKA (NUU-CHA-NULTH)
COAST SALISH RACING CANOE
COAST SALISH
COAST SALISH STYLE
SALISH SHOVEL-NOSE RIVER CANOE
Puget Sound
CHINOOK
WASHINGTON
Columbia River
OREGON
YUROK RIVER CANOE
YUROK
CALIFORNIA

CARVING THE DUGOUT

There are a variety of accounts how canoes were carved, though the art itself all but disappeared along with other losses of Native ways this last century. As Indian children were required to leave their families and attend government boarding schools, traditional ways and languages fell into disuse. Countless generations of accumulated knowledge went by the wayside in a push to modernize.

With the introduction of other modes for travel and the outboard engine in the 1920s, the beautiful old fishing and traveling canoes were no longer economically essential nor even viable.

Many dugouts used for fishing at accustomed sites were seized by government officials until the upholding of Indian fishing rights in the late 1970s. Forestry practices and clear cutting left few old growth cedar stands available for the remaining canoe carvers.

Many Northwest Coast men and women know how to carve though only a handful can make a good canoe. The tools and tricks of the trade are jealously guarded secrets. The ability to make a fine canoe continues to be one of the most respected skills.

Ancient ones found the inner strength and confidence necessary for such an undertaking through family inheritance and guardian spirit quests. One guardian spirit power, "Yeyi-ha'dad," was obtained from cedar trees for canoe making.

Elmendorf, an anthropologist working in 1960, recorded the knowledge of Henry Allen, one of the last Twana carvers. "The tree talks to you in a vision, gives you power to be a canoe maker. My grandfather Sustx had it." Sons could sing the power songs of their fathers though preferably a young man dreamt his own songs.

The cedar tree has special names. The male or coarse grained cedar is called "palowqwtan," the fine grained or female is called "paloqwtanot." When those possessing power or "su'win" felled the trees, they called them by their names to tell them which way to fall so they would not break. Only the persons who knew "su'win" knew these names.

Boas recorded a prayer for the falling moment: "Do not fall too heavily, else you, great supernatural one, might break on the ground."

Alone and out of sight of others, the carver might receive mystic help. If properly sanctioned, as he approached his carving spot early in the morning he would hear the rhythmic "chop-chop" of

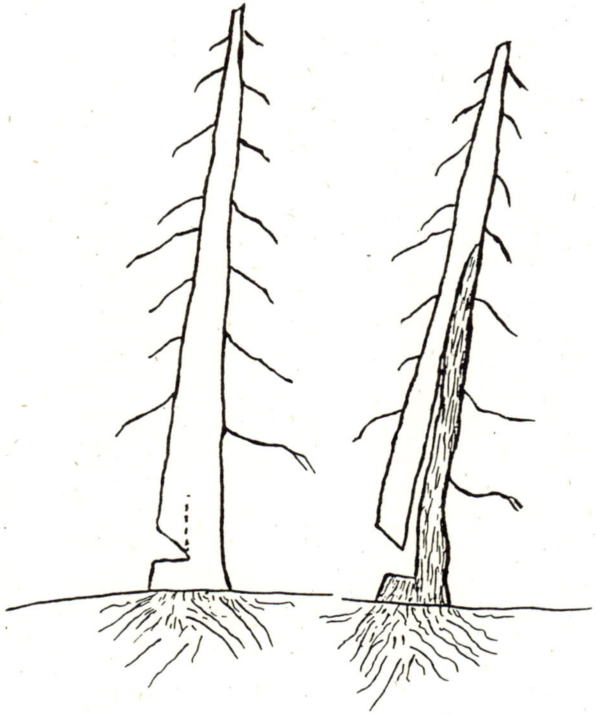

Stone chisels, adzes and yew wood or deer horn wedges were inserted into the tree to initiate the felling. Drawing by Chief Jack Khahtsahlano (Kitsalano), 1943, #37 Vancouver City Archives.

The Port Gamble S'Klallam felled this old growth cedar from the U.S. National Forests for their 35 foot centennial canoe. Gerry Kearney photo, 1987.

a supernatural assistant working on his canoe. The woodpecker, especially the northern pileated redhead, was the most common dream spirit of canoe builders among the British Columbia Salish. The canoe maker who dreamt of it was certain for success in his profession.

Ritual purification required that a canoe carver not comb his hair for fear of splitting the hull and that he also practice abstinence. This ensured the clear energies of the carver. It has also been suggested that the required abstinence would encourage the carver to more quickly finish his canoe.

A sound red cedar with few branches is rare and essential. Drift logs found along the beach were valued for potential canoes, though the more common source were trees selected in the forest.

After a cedar tree of suitable size with clear, almost straight grain was found in an appropriate location, the tree was tested to determine the soundness of the inner wood with a chisel. This was "to feel inside" the belly of the selected tree. Old growth cedar has the tendency to be soft in the middle. For strength in the bow and stern portions, this rot is avoided.

In ancient times canoe trees were felled by chiseling or controlled burning. Chief and carver, August Jack Khahtsahlano, was asked if cedar slabs were cut from standing trees. He indicated that the wood for a canoe could be felled by undercutting half the base of a cedar tree with a stone chisel and then inserting yew or deer horn wedges upwards. The wedges would cause the half-tree slab section to split upwards and proceed to fall with a minimum of effort.

For moderate sized Salish canoes, only one half of the log was used. The split surface became the top of the canoe. For large craft a whole section of a 400 to 800 year old cedar tree was needed. In spots rich in quality timber, the carver might fall several trees at once.

By combined labor, numerous logs were gotten out at one time. In the 1900s the felled logs were made into a raft, and by means of sweeps and sails and by dint of working the tides were brought to a village or neighborhood camp.

In other cases the splitting, shaping and hollowing of the hull took place on the spot of felling, often at some distance from the village. When the rough-hewn shell had been made ready, the carver asked a few friends and relatives to help

Iron adze blades and other metal tools were on the coast long before the European explorers, from drifts off Asian shipwrecks and overland trade.

get it to the water and tow it to a beach close to home. There he did the final shaping.

Green cedar is more easily worked than drier, older wood. While still thick the log is more vulnerable to deep checking. Women continue to strip bark off the felled cedar and weave it into mats, baskets, capes, skirts, cordage or other useful items. The length of the hull is determined by cutting the bow and stern ends. Boas recorded that the butt end of the log was often a wider section and used in the bow.

Salish carvers use the D-adze, so named for the handle's characteristic "D" shape. The adze is wielded like a short handled axe except that the blade of the adze is at right angles to the swinging motion. Careful splitting with elk horn or yew wood wedges and stone hand mauls once removed much excess wood. A fire controlled with clay also burned off sections of wood. A bone auger was used for a drill.

In ancient times jadeite and nephrite blades shaped the canoe. Mussel shell knives and crooked knives made from beaver's incisors were handy. Highly valued iron was found on shipwrecks carried by the current from Asia and also traded overland from European contacts.

Makah, Quileute and most Coast Salish carvers used the stone maul and D-adze. Museum of History and Industry #9213.

In modern times the chainsaw successfully roughs out the dugout hull, yet the adze is such an effective device, carvers still use it as their tool of choice.

The side with the fewest branches is chosen for the hull's bottom. Kwagiulth canoe and fish boat builder, Ernie Scow, suggested another method. If located near the shore, the log is put in the water and usually settles with one side downward. This side is selected to be the bottom of the future canoe.

Before beginning to rough out the hull, carvers must consider the changes which will occur during steam spreading. Native canoe makers devised a way to widen the beam or cross-width of the canoe beyond the diameter of the original log. For example, if a cedar log was only 3 feet wide, they were able to

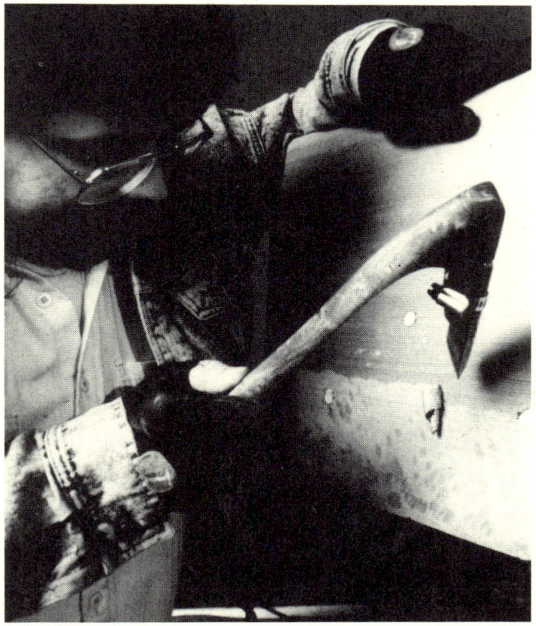

Joe Gobins, a Tulalip carver, uses the elbow adze to cut a fine finish. David Current photo, 1989, Suquamish Tribal Archives.

increase the canoe's width to 4 feet with steam spreading.

More specifically, the figures for a few recent replicas are: Holm's 35 foot Northern style canoe was spread from 42 to 60 inches—an increase of 43% of the original beam. Bill Reid's 50 foot Haida canoe, the *LooTaas* or *Wave Eater,* was spread from 54 to 73 inches—an increase of 33%.

Canoe makers envision which parts of the hull will change—particularly the gunwale shape, flare of the sides and keel rocker; and which parts of the hull will remain essentially stable—particularly the bow, stern and curves at the turn of the bilge.

Today's experienced carvers recommend that the best preparation to foresee the changes is to first carve and spread a scale model.

Carving begins by removing the sapwood whenever possible. The top surface is split and adzed level. Onto this, the canoe's top view shape (before spreading) is determined along the gunwales from bow to stern. This shape is roughed out. Next the log is rolled over, bottom-side-up, to contour the hull. Both sides are carved alike from end to end. Quality workmanship is easily recognized in the regularity of adzing.

While upside down, a nicely calculated "hog" or upward curve is given to the keel line, fore and aft. This reverse camber will change later in the widening process. The sides of the unspread canoe are left parallel sided. The bottom of the hull through the waterlines 3 inches above the keel are crucial and must be carved in a fair curve—not left flat like the higher sides, because the hull changes little at the turn of the bilge during spreading.

Once the unspread shape of the outside of the hull has been neatly adzed, it is ready to turn over and be hollowed out. Controlled burning was one ancient way to accomplish this. Burning was probably done only in strategic places and sections split out between burns. Another technique for excavating this large quantity of cedar is to carefully split off sections using wedges and mauls. Today, canoes are often roughed out with a chainsaw.

The challenge to preserve a uniform thickness along the side was cleverly solved. Twirling his awl skillfully between his palms, the ancient craftsman drilled rows of tiny holes at regular intervals. Maple or yellow cedar plugs cut to specified lengths were set into the holes from the outside as guiding depth gauges. The canoe maker could then cautiously adze out the inside of the hull until the measuring plug appeared, marking the ideal finished thickness.

In order to spread without splitting, the hull must not be too thick. If the turn of the bilge is not thin enough, it will not spread. Holm measured the fine older canoes kept in the Smithsonian and notes they have extraordinarily thin hulls.

BEFORE & AFTER SPREADING A COAST SALISH CANOE

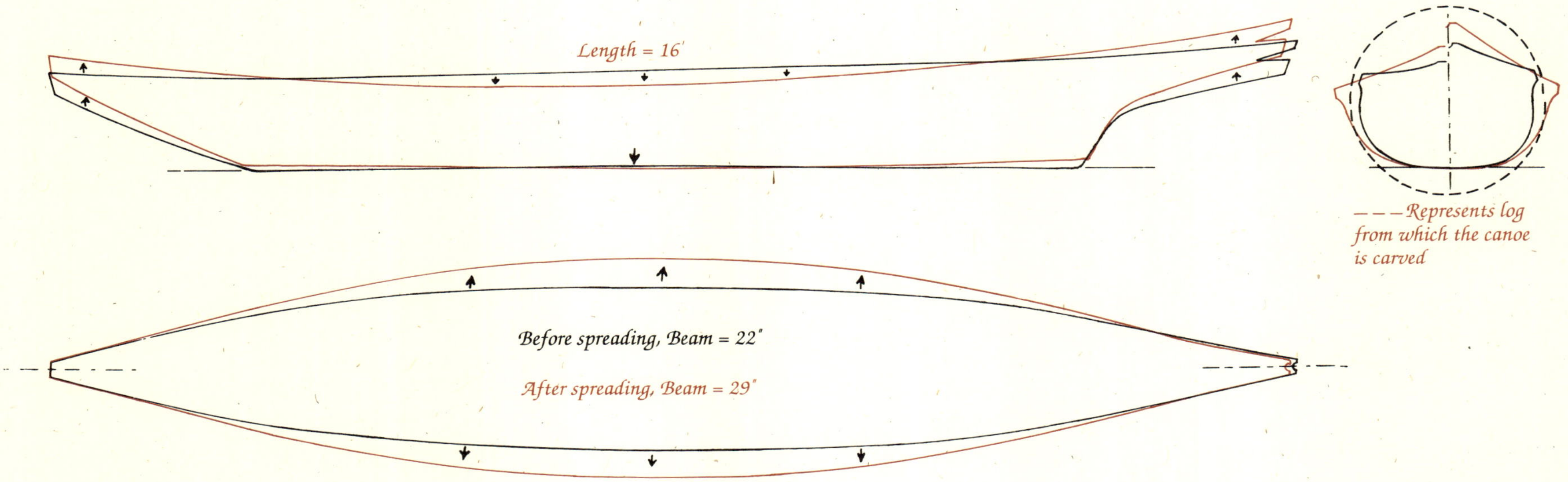

Built by Bill Holm, 1990. Lines taken by Holm and drawn by Leslie Lincoln, 1992.

Midsized 25 foot canoes are one finger thick (usually 3/4 inch) along the sides, gradually thickening along the turn of the bilge to two fingers thick (1-1/2 inches) on the bottom. The 58 foot long by 8 foot wide Nootkan canoe at the Smithsonian is only 1-1/2 inches thick along the sides. A 44 foot Haida canoe is only 3/4 inch thick below the gunwale groove and only 1-1/4 inches at the turn of the bilge. The slightly thicker bottom adds stability and strength to the boat.

Spreading is the critical point of the whole job. If the carver works too fast or is careless in this operation, he might split or break the sides. In order to produce correct lines in the finished craft, the spreading must be just right at every point.

At dawn on the spreading day, the day the ungainly carved log achieves its graceful form, the carver and his trusted helpers build a fire not far from the side of the hull. They place round stones in

The carved log is transformed into a graceful canoe by the steam-spreading process.

with the coals and heat them until red hot. Water is poured into the canoe shell approximately four fingers deep, just enough to cover the rocks and generate the penetrating steam necessary for flexing the wood fibers. Canoe makers sometimes use seaweed and stale urine inside the canoe to further soften the lignin which binds the cellulose fibers.

Green wood tongs are used to move hot stones into the canoe bath. This water is heated to a frothy boiling mixture. Cedar bark mats (or plastic sheets) cover the hull and help the heat penetrate. The carver splashes steamy water along the sides. This moist heat allows the wood to flex and the interior fibers to expand. The dry heat of the fire outside causes the outer fibers to dry and shrink. These differing forces encourage the sides to spread out.

Carefully, the knowing carver taps temporary thwart pieces across the hull in a slight fore-and-aft angle. As the hull softens he taps these thwarts to right angles, applying continuous pressure to widen the hull to its desired full shape.

The horseshoe-shaped sections flare out increasing the beam. The fore and aft humpbacked camber of the gunwales draws down into a graceful, sweeping sheer. Waterline curves become fair. The hog in the keel flattens out to a straight run and may develop a gentle rocker.

The canoe cannot be spread too far, nor too quickly. The added weight of the rocks and water within the canoe further assist the bottom to become level. The bow and stern rise slightly as the gunwales fall open.

When the canoe maker judges the shape to be just right, he removes the rocks and water. The canoe slowly cools into its new form. Later, any slight unfairness will be adzed off.

On the inside of the hull, in order to complement the shape and thickness of the characteristic outer gunwale rim, an inner groove is carved the length of the hull, usually 3 inches wide. Thwarts serving as seats in the bigger canoes, are set onto this groove. The Salish sewed permanent round thwarts into place using strong, flexible cedar withes. Thwarts also have been pegged into place. Boat nails and plastic coated electrical wire came into usage when they became available.

Usually a protective caprail of fir is added along the upper edge of the gunwale to save it from chafe caused by paddling. In the bottom of a smaller hunting canoe the paddler knelt on a sewn cattail or bullrush mat. Seats are set low in the hull to maintain stability. In recent bigger canoes, lead ballast or rocks placed in the bilge add stability.

Nooksack carvers once charred the outside of their canoes black to harden the cedar and bring the protective resin to the surface. The canoe was also coated with oil from the whale, seal or dogfish liver. Oil reduces friction between the boat and water, perhaps increasing speed. It also seals the grain. Some applied red paint—a mixture of colored clay ochre and salmon roe, inside the canoe or simply along the inner gunwale groove.

In 1986 the progressive Haida employed a low-tech/high-tech plan. They synthesized modern with ancient technology while rediscovering their art of canoe crafting. In preparation for a 1987 "Voyage Home" from Vancouver, B.C., to the Queen Charlotte Islands, craftsmen repaired cracks in the 50 foot *LooTaas* with oak and stainless steel ribs, all saturated with five coats of epoxy sealer.

In simpler times, the dugouts became water-logged and due to beaching became frayed or feathered. So periodically canoers thoroughly dried and then either lightly charred or carefully adzed and polished their vessels to remove a thin layer of the worn outer wood.

Cedar dugouts are unsinkable but they are susceptible to checking and splitting. If exposed to direct sunlight, splits develop along the length of the hull. Knowing hands repaired such splits by lashing them with cedar withes. Wedges inserted into the lashing ensured a tight repair. For caulking they used shredded inner cedar bark and pitch. If one looks at most older canoes, one will see scarfed pieces and tin patches. With frames to reinforce against splitting, canoes can remain in service for more than 70 years.

Native people understand the belief that spiritual knowledge adds strength. Stern, in 1934, working with the Lummi recorded that on a stormy day, if there was danger while traveling the man who understood "su'win" slapped the sides of his canoe and shouted, "Palo-qwtan take us home safely. Wake up! Help us through this." The canoe responded to his words, came to life and carried the voyagers safely through the storm.

Splits in the 50' Haida canoe were stabilized with frames and sealed with epoxy—a progressive low-tech/high-tech watercraft. Photo by the author, 1987.

For centuries splits were caulked with shredded cedar bark and pitch, then lashed with cedar root withes and tightened with wedges. Newcombe photo, 1913, Royal British Columbia Museum PN 629.

PADDLES & SAILS

Quick handling and perfect timing are essential for the race crews. David Current photo, Suquamish Tribal Archives WB6B #30.

SCALE DRAWING OF TULALIP PADDLE *By the author, 1988*

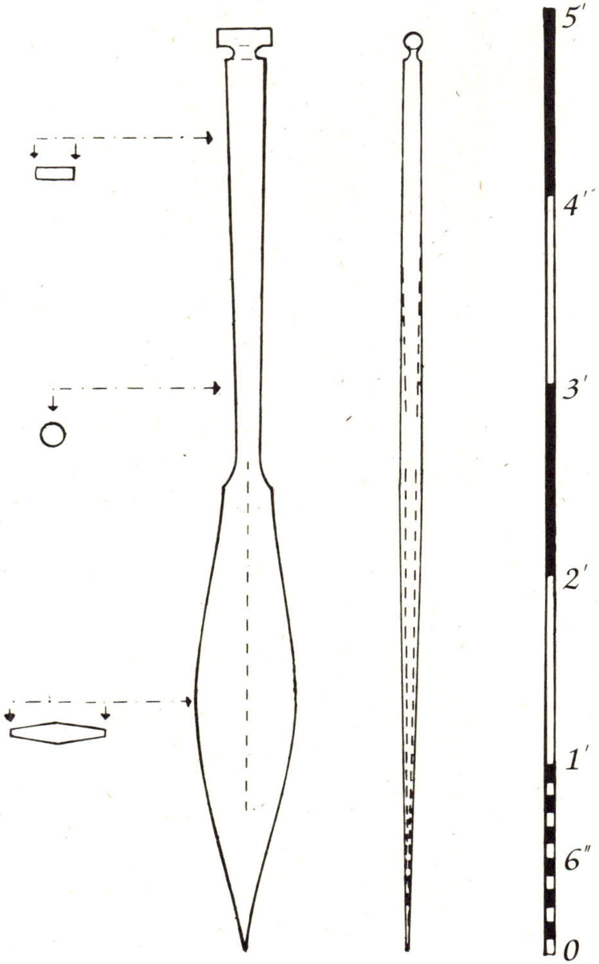

Paddle types vary, though all are of the same general shape, carved of yew, vine maple, red and yellow cedar.

The blade of the Coast Salish paddle is leaf-shaped, thin along the blade and pointed at the tip. The handle is a cross piece usually carved of the same piece of wood.

Swan was amazed by Native seamanship, particularly a crew's ability to use paddles to right and steady themselves when severely heeled by wind or wave.

In historic times, the more common types of paddles included the man's hunting paddle with a sleek diamond blade and particularly fine point. The freighting or general traveling paddle was heftily built for extended use. The woman's paddle was shorter and carved with a slightly fuller blade, some almost rounded.

Coast Salish used a folded bark bailer. The Twana favored a "sugar spoon" bailer and the Makah carved theirs in rectangular form. Bill Holm slide.

The paddle shaft varies among tribes. The shaft of a Twana paddle is like a rod and kelp was sometimes wrapped around this for a comfortable handhold. The Tulalip paddle shaft, as well as paddles to the north, are rounded just above the blade and rectangular in cross section just under the upper handhold. The racing paddle is lighter, especially under the handle.

Some say the paddle's pointed end was designed for warfare. Although the point could be a weapon and was handy to stab a flounder, the tip of the blade is actually a fine-tuned innovation. The tip enters the water's surface more easily than the blunt-tipped paddle. When silence is required, the very end of the tip can be left in the water on the return stroke so that even the slightest noise of drips splashing the surface is not made.

The Chinook or "river-poling" paddle is the most unusual kind. Its blade is shaped on the lower end as a crescent rather that coming to a point. This shape is useful for making way against the current by catching onto roots and into the mud. However, Steve Philipp warned these crescent paddles tend to split lengthwise in the center of their concave end.

River canoes were propelled by fir poles, which were used to push the canoe upriver. The saying goes: "You push your pole and pull your paddle."

Paddles are made with pride. Many are painted on their blades though the southern paddles were frequently left unadorned or colored solid black. Waterman recorded one method for blackening the paddle. They used fungus found on hemlock bark as a pigment. These fungi were boiled, then applied as a stain. When the stained blade was held over a fire it turned glossy black. Attentively holding the blade over a pitch fire is a simpler blackening technique. Modern paints are now used on racing paddles. Unique designs acclaim the crew's identity.

Some carvers use the reach of their arms to determine the proper length for a paddle. Others described the length as the distance from the paddler's chin to the ground.

Native pullers treat their paddles with care and utmost respect. They explain the tip should never touch the ground but can rest on one's foot. The tip is sacred as it is the meeting place between the puller and the water. The paddle is powerful because it helps you to get from here to there and to reach your goals.

Photographs from the turn of the century reveal that Native mariners appreciated oars. Canoes up and down the coast were fitted with rowlocks and propelled by oars in addition to a person at the stern with a steering paddle.

Water is splashed from inside canoes with a paddle though bailers are more effective. Most common in the Straits and on the Sound were bailers folded from cedar bark, similar in construction to makeshift bark hunting canoes. The base was a rounded section of bark and the sides were a continuation of the inner bark pleated up at right

Wooden slats of cedar were lashed together to serve as an aboriginal sail.

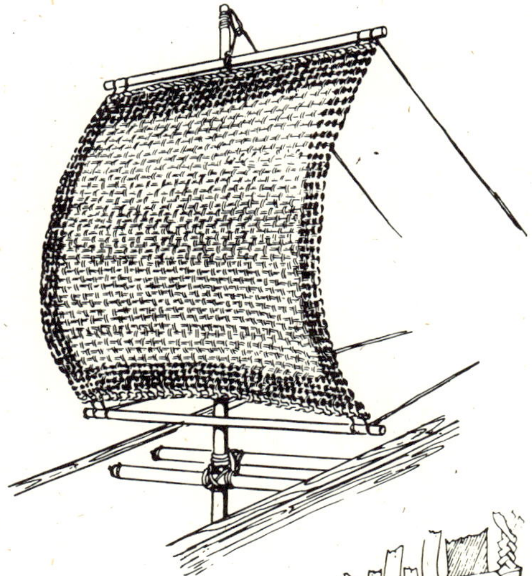

ABOVE—The Makah used a woven cedar bark square sail raised up the mast with a halyard of twisted sea lion gut. BELOW—Detail border weave of the clew or lower corner of the bark sail. Sara Vinsonhaler drawings, first published in The Whaling Equipment of the Makah Indians, University of Washington Publications in Anthropology by T.T. Waterman, 1920: 23-24.

angles and lashed with cherry bark twine to a cedar wood handle.

Salish men and women did not always paddle. If conditions were right they were happy to move with the force of an accommodating wind. They used sails made of woven cattail, rush, cedar bark and even of sewn slats of wood. The small sails could catch a favorable following breeze.

Edward Curtis described the aboriginal Coast Salish sail as a strip of rush matting, supported on two yards lashed at their middle to the mast. In order to take in the woven sail, it was necessary to unstep the mast. The mast was held fast by slipping it through a strong thwart and stepping it on the bottom of the hull.

Boas stated that the woodslat sail was the type Native peoples used before contact. He described the lashed cedar board sail which was thinly split and sewn together—a total surface area of 12 square feet for use on a 35 foot traveling canoe.

> *Now that is the sail of the ancient people before any white people came; to wit, short boards sewed together. The canoe-mast is short, for it just shows above the top edge of the board sail when it is standing up in the bow. The wind just*

ABOVE—This unlikely looking sail was made of strong, thin slats of cedar, useful for sailing downwind. Royal British Columbia Museum PN 2057.

Indigenous seafarers adapted sprit-rig cloth sails for use after the arrival of sailing ships. McCurdy photo, circa 1890-1910, Museum of History and Industry.

> *blows against it and presses the board sail against the mast when the canoe is running before the wind (Boas in Howay 1941).*

The Quinault also used wooden sails for downwind propulsion. In 1936 Olson noted that they made no attempt to tack with a side wind. The sail was always carried at right angles to the boat. If the wind were not blowing from the stern, the sail was taken down.

Maritime historian Frederick Howay found no accounts recorded by European explorers of canoes rigged with sails. The first observed use of the cloth sails are of the sprit-rigged type like the sails of the ships' longboats. Bartering for sail cloth and sailing rigs was noted as a prominent trade interest.

It seems highly probable that before contact, the adept and resourceful aboriginal mariners crafted sails and used woven mats to aid them in downwind runs.

FISHING & HUNTING

This fishing weir built across the Cowichan River (circa 1864) was carefully checked for mink and weasels attempting to steal the valued salmon catch. Dally photo, Royal British Columbia Museum PN 1380.

Fishing was and continues to be the basis of the Coast Salish economy. The dugouts were essential for such a livelihood. Many species of fish were caught. Clams, geoducks, mussels, oysters, crab, abalone, scallops, barnacles, octopus, sea urchins and many types of seaweeds were gathered along the intertidal shoreline. Seal and porpoise were hunted in the sleek Coast Salish canoes. Hunters caught waterfowl, geese and ducks from canoes in the night and by flyway nets strung up along flight routes.

The Sound and Straits were rich with resources. The rivers and estuaries were alive with marine creatures which could be hunted, fished or gathered in their own time. From Ballard's and Waterman's work, David Buerge has described how the Duwamish people—the first inhabitants of the Seattle area, traveled. These Salish were a mobile, waterborne society following seasonal runs to set up camps at well established sites.

Whirling flocks of screaming gulls marked schools of herring ready to spawn on eel grass in the Sound during the February and March runs. Natives in canoes followed the herring balls, dipping their herring rakes into the masses. The rake was a paddle-like implement, inset with bone barbs like a comb. The rake was stroked like a paddle in a sweeping motion, spearing the herring and then depositing the little fish into the canoe in one clean stroke. Smelt were also fished in such a manner or with nets.

In April and May fishing camps were set up at the head of Elliott Bay prior to the spawning run of Chinook. Relatives from other groups came and there were many mat lodges and drying racks set up along the shore. Scores of canoes trolled the bay. Farther out, hunters in their lithe craft went after the seals and blackfish that preyed on the converging salmon.

Gradually this activity worked its way upriver when the runs commenced and the camps swelled around weir sites. Traps and weirs were constructed across rivers to catch salmon during their peak runs. Dipnets and harpoons were wielded from stable canoes. These were lashed to the platforms while fishermen tended the weir.

MAP III

Ancient Seattle

WATERS OF THE DUWAMISH

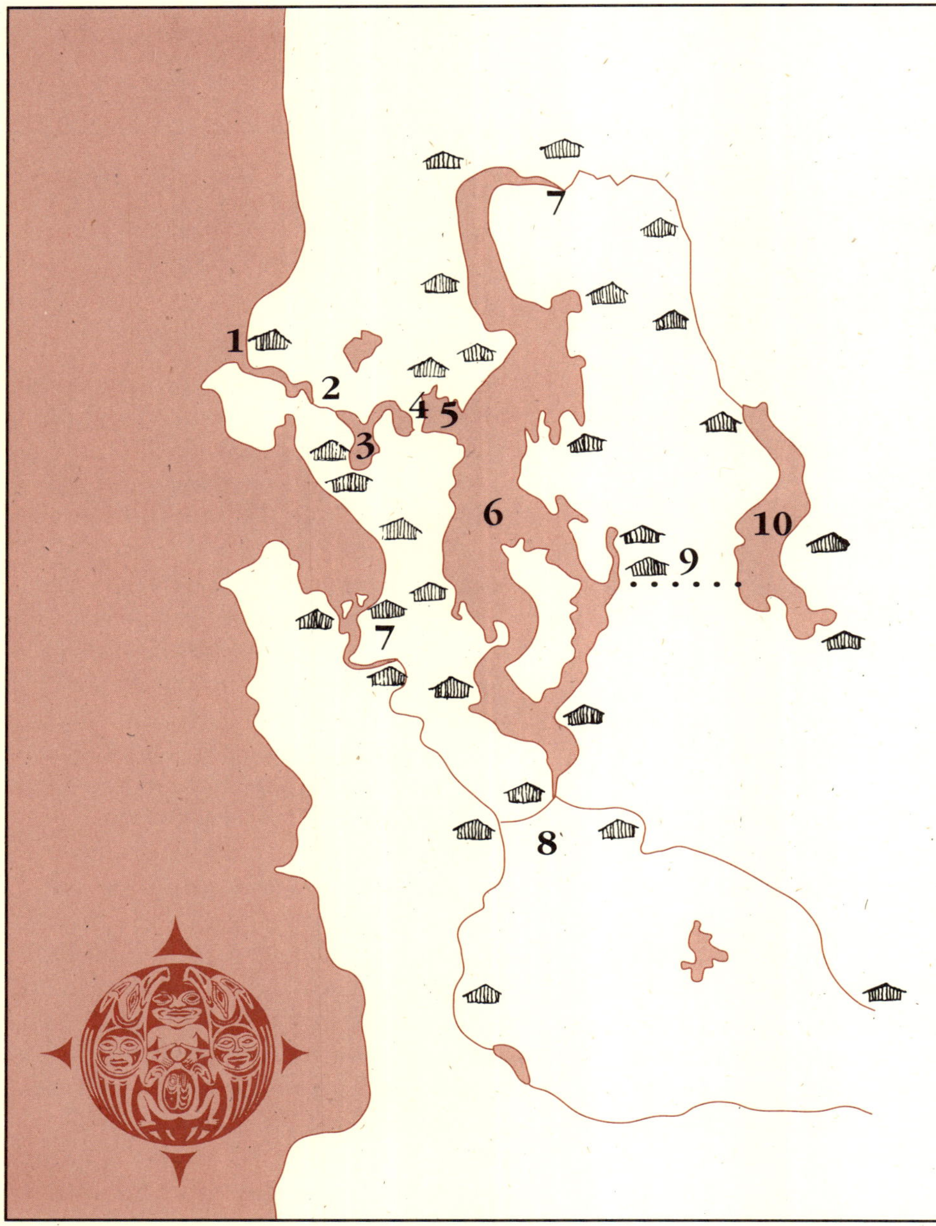

1. SHILSHOLE in Lushootseed, the Puget Sound Salish language, means "To thread a bead" in reference to the marsh entrance.
2. Saltwater peoples meandered through the Shilshole estuary catching the tide into Ross Creek as far as present day Fremont Bridge.
3. HA-AH-CHU or Littlest Lake (also TENAS CHUCK in Chinook Jargon) now known as Lake Union.
4. JIJILA' LETCH at the portage meant "Little crossing place." It was a protected crossroads and important trading center for saltwater, freshwater and inland peoples.
5. SHUL-WEHL referred to narrow marsh passages. The term originally referred to tiny holes bored into a canoe's hull to measure its thickness.
6. HAHT-HAA CHU-ABSH means Big Lake (HYAS CHUK in Chinook) now known as Lake Washington. Before the locks the lake drained south, flowing through the Black into the Duwamish River.
7. Seasonal camps were near fish weirs or traps set up across rivers and creek mouths.
8. MOCKS LA PUSH (or Black River) meant "River with two mouths" as it flowed into the lake with seasonal flooding.
9. Lake John's Village and Trail.
10. HAHCH-HAH-CHU, "Second, smaller lake," now called Lake Sammamish.

NOTE: These pre-1850's house sites are marked in approximate locations (Buerge 1984; Waterman 1922; Chrzastowski 1983).

In the 1850s an early pioneer, Dr. David Maynard, asked Chief Sealth, after whom the city of Seattle was named, to show him a good location for catching salmon commercially. Sealth took him to the mouth of the Duwamish River. Maynard hired Indian fishermen and the catches were salted and transported to San Francisco. For many years as many as 60 canoes at a time could be seen in Elliott Bay.

For millennia before Seattle's waterways were rerouted by opening the Chittenden Locks in 1916, annual migrations of salmon swam up the Duwamish, Cedar and Black Rivers to enter the south end of Lake Washington. Fishing weirs near the confluence of the Cedar River and at the lake's outlet were of considerable importance.

Until 1916 Lake Union and Lake Washington were separated by the Montlake divide or *JIJILA' LETCH*—now Portage Cut. This had been a gathering place for the freshwater, saltwater and inland peoples. Lake Union drained through a creek into Shilshole Bay. Lake Washington's famed land-locked sockeye called "kokanee," drew fishing parties from all around the neighboring tribes.

Other important portage sites were to be found between significant bodies of water. The S'Klallam and earlier the Chemakum hosted a protected portage site named *TSESTIBUS* or "Pull canoes across" by Indian Island near Port Townsend.

Before 1900 Commencement Bay was a popular gathering place. Gambling bone games lasted long into the night. Museum of History and Industry 3.488.

The Chehalis and ocean-dwellers could pole up river almost to Puget Sound, crossing over a bit of land south of what is now Olympia. The entire territory was known for its trade routes, clamming beaches, berry and camas patches and strategic portage locations.

Hop-picking was a popular industry in the late 1800s. As the hops ripened, canoe crews and their families traveled toward fields in the Snoqualmie and White River valleys. They followed established routes, set up familiar encampments and enjoyed family reunions.

Halibut, cod and dogfish were longlined in deeper waters. Primarily the women gathered, dried and smoked the shellfish. Harbor seals and porpoises were taken. Whaling, as practiced by the Makah ocean-dwellers, was not done by the Coast Salish because whales rarely entered their territorial waters.

Sea mammal hunting was prestigious work. Skilled hunters brought home seals, sea otters, sea lions and porpoise. Porpoise hunting involved a two-man team. The harpooner was seated in the bow and the steersman in the stern. A paddle-rest was made by wrapping soft cottonwood bark around the thwart. Upon approaching their prey, the harpooner silently laid down his paddle, stood up, took aim and drove the harpoon to his kill.

Makah hunters harpooned migrating grey whales. A. Curtis photo, circa 1900, Special Collections, University of Washington Libraries NA 740.

The 15 foot harpoon was made of fir with a foreshaft of ironwood. The harpoon maker lashed together a composite point of bone and barbed elkhorn with cherry bark wrapping and pitch. Inland Salish neighbors traded hemp which was made into strong leader cord for the harpoon lines. Floats were sometimes used for porpoise hunting. The hunter dispatched his prey with a short spear, bone or stone club. Elmendorf records a seal was brought inboard while the hunters towed porpoise to shore.

Waterfowl hunting occurred in the spring and autumn when migrating fowl landed in the lowland estuaries, marshes and sloughs. Two man teams caught geese, brant, bluebills, golden-eye, canvas-back, teal and mallard species. They snared the birds in their Sda'kwihl canoes with spears or a "pole net"—a square sinew or nettle-fiber net attached to the end of a pole.

Elmendorf describes the Twana's night hunting technique. Their clever strategy made use of a "fire and shadow screen" to herd fowl into snaring range. The hunters built a fire in the very stern of the canoe on a small earth-covered platform and fed it with silently burning pitchwood. The paddler in the stern wore a special woven headdress held on a stiff frame down to his shoulders. As he faced forward his headdress cast a shadow over the rest of the canoe. By moving his head he could direct the glow of the firelight. Waterfowl that swam out of the evening darkness into the firelight were frightened and returned to the shadows. A flock could thus be bunched, or a single bird herded in front of the canoe.

Hunters learned to travel in silence. They greased their paddles to ensure imperceptible movement through the night waters. The stern paddler indicated the exact moment to capture the water birds with a slight rocking of the canoe. No words were spoken. The person in the bow had no difficulty capturing the fowl with spear or net.

Duck hunting changed with the introduction of the gun. No longer was the quiet fire and shadow technique appropriate. Lawerence Webster, an elder from Suquamish, recalled the first time he used a gun to hunt fowl. He aimed and fired across the gunwale. The recoil upset the canoe which nearly tipped over! The noise scared off all other prey nearby and his uncle's gun went overboard.

Suttles describes how the Straits Salish fishing folks developed an effective way to catch the highly valued sockeye. They watched the salmon swim along naturally funneled routes among the reefs. Two canoes worked as a pair. Together they anchored over a specific location, usually near shore on a kelp-covered reef that lay in the sockeye's path.

The keen fishermen strung a darkly dyed net between their canoes. Following familiar migration routes, the sockeye swam into the net. The captain standing on a raised platform sighted the fish. The crew drew the net up and deftly landed the sockeye. This unusual technique

The Straits Salish people invented reef-net fishing, a unique strategy still used today.

ABOVE—These 1930's Lummi reef-net fishermen caught sockeye in a net held between their anchored dugouts. E. Field photo, Special Collections, University of Washington Libraries NA 1804.

LEFT—Pairs of carvel planked reef-net fishing boats caught Straits sockeye in the 1940s. Wayne Suttle photo, Thomas Burke Memorial Washington State Museum #L3911/2.

created a fine opportunity to harvest the salmon before their upstream spawning journey.

Suttles' informants explained that in the 19th century the reef-net canoes were of a special type: large with a wide bow and flat stern somewhat similar to the shovel-nose. They were safely used in the Straits because the weather had to be calm for this fishery in order to see the salmon swim over the net. Fishermen replaced the early dugouts with Nootkan style craft and these were subsequently replaced in the 1930s by plank-built skiffs.

The fine yields of this fishery have encouraged Indians and whites alike to still reef-net migrating sockeye in the San Juans and Straits areas though the cedar dugouts have been replaced by stable aluminum scows.

The seasonal travels of the Coast Salish slowed in the autumn after the salmon runs. People returned to their winter villages, their canoes laden with tons of smoked and dried salmon, other winter provisions and newly bartered goods. By late November most canoes had been pulled up on skids above the high-tide line. In Lushootseed this was "*SHSCHEE-CHUL-WAHS,*" the time to put your paddles away and keep close to home.

CANOES IN CEREMONY

Because of its critical role in the Native lifestyle, the canoe also plays a starring role in dances, stories and in song. Paddlers often sing while under-way to keep everyone's mind as one and the paddles in time. Paddling songs are also sung at certain gatherings to honor crews and past voyages.

On special occasions or when arriving from some distance, a lead dancer will perform in the bow announcing to the assembled group who his or her people are. Paddle handles are beat against the hull which resounds with the excitement, drums and voices heard from ashore.

Canoe launchings, dedications and naming ceremonies are a joyous time to recognize the carvers. All witness repayments of debts to those involved who secured and felled the tree, who fed the carvers and who helped carry the craft to the water. It is a time to bless the vessel and crew for future journeys.

An ancient winter renewal ritual of the Puget Sound Salish heritage was the spirit canoe ceremony called *SBE-TETDA'Q*. It was banned along with most traditional Indian religious practices by missionaries. Yet Waterman recorded a reenactment of it on the western shore of Lake Sammamish in the early 1920s.

Shamen or trained healing doctors performed the spirit canoe ceremony in a village dance house during the winter solstice. This was a particularly dark and difficult time when lonely ghosts might kidnap one's soul. Healers were hired to make a ritual journey to the "Land of the Dead" in a spirit canoe. If successful, they retrieved the lost soul during battle with the ghosts and returned it to the patient.

Six carved and painted cedar planks, each 3 feet high, were planted upright in two rows in the earth to represent the spirit canoe. The bent-over tops of these

At the PADDLE TO SEATTLE landing ceremony on Shilshole Beach, Mandi Jones danced in the bow of the Port Gamble S'Klallam canoe. She spread feathered down as a sign of peace. Ron Peltier photo, July 1989.

Life without a canoe was unfathomable. Even in death the wealthy were placed in one to travel to the Land of the Dead. Though canoe coffins are no longer used, two monumental canoes honor Chief Sealth's grave in Suquamish. Author's photo, 1984.

spirit canoe boards faced the eastern dawn so the powers portrayed by them could "swallow the daylight," enabling the healers to safely return in the morning. One plank represented a cetacean with flukes, flippers, tongue, teeth and dorsal fin. This mythical monster, *SKEBA'KST* drew people into its maw by sucking its breath. Dots painted on the boards represented the songs sung.

Into the night all sang as the traveling party went deeper along the ghost river. When the lead shaman finally located the missing soul, the voyagers turned their vessel around by interchanging the canoe planks at either end. After pushing off, the captain in the stern hurled his "meanness" or power into the ghost village. A furious skirmish followed with showers of arrows as the spirit canoe quickly escaped downriver with the retrieved soul aboard.

Each shaman carried a staff which served as a paddle, a pole for moving the canoe over shallows, a spear and as a bow. Carved posts or images of spirit guardian figures 3 to 4 feet high brought strength and guidance for the journey into another realm.

Before the arrival of Euro-Americans, Native people often laid their dead in canoes supported by poles located on islands or in the branches of trees. They headed these burial canoes downriver toward the sea. This reflects an intimate connection between a mariner and his or her vessel. Shrouded within the protection of one's personal canoe, the deceased's soul could journey to the Land of the Dead. Olson describes this type of interment:

> *Probably the commonest [burial] method was in a canoe. One or more holes were cut in the upright canoe, the custom was interpreted as signifying that the canoes used in the Land of the Dead were so perforated. A related idea was that the canoe so treated was "killed" and went to the other world where the deceased might use it.*

A RESURGENCE

Paddles up on the Port Gamble S'Klallam canoes. Jake Jones photo, 1989.

After a hiatus of several generations Natives in Washington began again to carve their traditional canoes. With sponsorship from the Washington State Centennial Native Canoe Project, more than 17 tribes became involved in rediscovering the art of making cedar voyaging canoes. The United States National Forest Service recognized the tribes' right for red cedar, thus carvers regained access to the valued old growth stands.

Thousands of Native Indian people participated by carving, restoring, racing, voyaging and welcoming back traditional craft at a splendid voyage known as the **Paddle to Seattle**. The paddle was a cultural pilgrimage involving many communities during a score of tribal encampments.

The Quileutes challenged other tribal nations to join a 340-mile round-trip voyage. The Bella Bella Heiltsuk paddled their Northern style canoe south from Canada, uniting Natives across the international border. Together they declared their continuing vitality in the contemporary world as they retraced ancient coastal routes.

A great flotilla of more than thirty voyaging and racing dugouts converged at Suquamish for a momentous encampment. The flotilla paddled to Seattle in Duwamish territory. As the cedar canoes once again plied Shilshole Bay off the ancient village site, the assembled tribal nations signed a proclamation declaring Washington state has come of age with its hundredth anniversary requiring all its citizens to become responsible stewards of this land. This same season the tribal nations affirmed their sovereign status vis-a-vis the Washington government.

Widespread enthusiasm for this waterborne movement continues. By meeting the challenges to relearn lost ways, to bring back the valued dugouts and to develop endurance and skill in the youth pulling crews, Native communities are strengthening a pride in the traditions rooted in their past. The canoe has become both a provocative focus for cultural renewal and a positive emblem for Native coastal identity.

Native eloquence best expresses the story behind this canoeing revival. The Bella Bella Heiltsuk people from the central British Columbia coast have challenged aboriginal canoeing peoples from around the Pacific to another magnificent voyage.

GAATUWAS

"People gathering together in one place"

A stroke toward indigenous unity — Beginning at Bella Bella, B.C., Canada, 1993

HYLBUS
"Magical Paddle"

The GLWA or the ocean-going canoe
Is what binds us all together
Because we are an indigenous
Maritime culture.
The canoe has sustained us
For the last 10,000 years
On the central coast.

What I've learned over the last five years
Is that if we can't learn
From our old people,
We have to go out and do it.
And we see why
Our ancestors did what they did.
And we take it and
We re-create our culture
Because we are a live,
Living indigenous culture.
We are one people tied together
Through our ocean-going canoe.

We invite all the tribes
To come to our village
Because what we've experienced
Over the last 470 miles
That we've paddled to get down here,
Over these last little while,
It was an Odyssey.

We've passed through
Over twenty tribes on our way down.
And we're challenging you
In the next stage.
In four years [1993]
We want you to come up to Bella Bella
And we're going to host
An encampment of ocean-going canoes.

Frank Brown,
Expedition Leader
Bella Bella Heiltsuk
July 20, 1989.

BIBLIOGRAPHY

Adney, Edwin and Howard Chapelle,
1964 *The Bark Canoes and Skin Boats of North America,* Smithsonian Institution, Washington, D.C.

Arima, Eugene,
1975 *A Report on a West Coast Whaling Canoe Reconstructed at Port Renfrew, B.C.,* History and Archaeology 5, Parks Canada, Ottawa.

Barnett, H.G.
1939 *Gulf of Georgia Salish,* Anthropological Records 1:5, University of California Press, Berkeley.

Boas, Franz,
1909 "The Kwakiutl of Vancouver Island," in *Memoirs of the American Museum of Natural History,* Vol. 8, Pt. 2, New York.

Buerge, David,
1983 "Salt Water Itinerary" in "The Seasonal Round of Activity," a manuscript of the Duwamish.

Carlson, Barry and Thom Hess,
1978 "Canoe Names in the Northwest, An Areal Study" in *Northwest Anthropological Research Notes,* Vol. 12, No. 1.

Clark, Ella,
1958 *Indian Legends of the Pacific Northwest,* University of California Press, Berkeley.

Collison, William,
1915 *In the Wake of the War Canoe,* (Republished 1981) Sono Nis Press, Victoria.

Curtis, E.S.
1913 *The North American Indian,* Vol. 9: Salishan Tribes of the Coast, the Chimakum, the Quiliute, and the Willapa, Limpton Press, Norwood.

De Laguna, Francis,
1972 *Under Mount Saint Elias,* Smithsonian Contributions to Anthropology, Washington, D.C.

Dewhirst, John,
1967 "The Salish Racing Canoe" B.A. Honors Thesis, University of British Columbia, Vancouver.

Drucker, Philip,
1950 *The Northwest Coast,* Cultural Element Distributions, 26, University of California Press, Berkeley.

Duff, Wilson,
1981 "Thoughts on the Nootka Canoe," (1964) reprinted in *The World is as Sharp as a Knife,* Abbot, editor. British Columbia Provincial Museum, Victoria.
1952 *The Upper Stalo Indians of the Fraser Valley, British Columbia,* British Columbia Provincial Museum, Memoir No. 1.

Durham, Bill,
1960 *Canoes and Kayaks of Western America,* Copper Canoe Press, Seattle.
1955 "Canoes from Cedar Logs, A Study of Early Types and Designs," in *Pacific Northwest Quarterly,* Vol. 46, No. 2: 33-39.

Eels, Myron,
1985 *Indians of Puget Sound, The Notebooks of Myron Eels,* George Castille, editor, University of Washington Press, Seattle.

Elmendorf, W.
1960 *The Structure of Twana Culture,* Washington State University Research Studies, Monograph Supplement, 2: 1-576.

Fladmark, Knut,
1979 "Routes: Alternative Migration Corridors for Early Man in North America," in *American Antiquity,* 44: 55-69.

Gould, H.
1935 "The Hydah Canoe" in *Alaska Sportsman,* Vol. 1, No. 3.

Gould, R.
1964 "Seagoing Canoes Among the Indians of Northwestern California" in *Ethnohistory,* 15: 11-42.

Haeberlin, Hermann and Erna Gunther,
1930 *Indians of Puget Sound,* University of Washington Press, Seattle.

Hamilton, Gordon,
1980 "War Canoe Races Still Excite Coast Indians" in *Canadian Geographic,* Vol. 2, No. 4: 38-45.

Henry, John Frazier,
1984 *Early Maritime Artists of the Pacific Northwest Coast, 1741-1841,* University of Washington Press, Seattle.

Heizer, Robert (vol. editor),
1978 *Handbook of North America Indians,* Vol. 8, "California," Smithsonian Institution, Washington, D.C.

Holm, Bill,
1991 "Historical Salish Canoes" in *A Time of Gathering, Native Heritage in Washington State,* Burke Museum, Seattle.
1987 "The Head Canoe" in *Faces, Voices and Dreams,* Peter Corey, editor, Alaska State Museum, Sitka.
1961 "Carving a Kwakiutl Canoe" in *The Beaver,* Summer, Volume: 28-35.

Howay, Frederick,
1941 "The First Use of the Sail by Indians of the Northwest Coast" in *American Neptune,* Vol. 1, No. 4: 374-80.

Kennedy, Dorothy and Randy Bouchard,
1982 *Sliammon Life, Sliammon Ways,* Talon Books, Vancouver.

Kirk, R. and R.D. Daugherty,
1978 *Exploring Washington Archaeology,* University of Washington Press, Seattle.

Lincoln, Leslie,
1990 Unpublished Master's Thesis *PADDLE TO SEATTLE, A Native Washington Movement to "Bring Them Canoes Back Home,"* University of British Columbia, Vancouver.
1989 *Native American Canoes, Washington State Centennial,* editor, Thomas Printing, Shelton.
1985 "The Quest for Native Canoes" in *Nor'westing,* Vol. XXI, No. 1, November: 32-39.

Niblack, Albert,
1890 "The Coast Indians of Southern Alaska and Northern British Columbia," in *U.S. National Museum Annual Report for 1888,* 225-386, Washington, D.C.

Olson, Ronald,
1936 "The Quinault Indians" in *University of Washington Publications in Anthropology*, Vol. 6, No. 1, Seattle.
1927 "Adze, Canoe, and House Types of the Northwest Coast" in *University of Washington Publications in Anthropology*, Seattle.

Palau, Mercedes,
1986 "The Spanish Presence on the Northwest Coast: Seagoing Expeditions (1774-1793)" in *To the Totem Shore*, Ministerios de Transportes, Turismo y Communicaciones, Madrid.

Powell, Jay and Vickie Jensen,
1976 *Quileute: An Introduction to the Indians of La Push*, University of Washington Press, Seattle.

Quileute Tribal School and Mark Masceri,
1989 *Paddle to Seattle*, a video production.

Roberts, Kenneth and Philip Shackleton,
1983 *The Canoe, A History of the Craft from Panama to the Arctic*, International Marine Publishing Company, Camden.

Smith, Marion,
1940 *The Puyallup-Nisqually*, Columbia University Contributions to Anthropology, 32, New York.

Stern, Bernhard,
1934 *The Lummi Indians of Northwest Washington*, Columbia University Press, New York.

Thirteen canoes in tow in Tulalip Bay, circa 1910. Brady photo, Special Collections, University of Washington Libraries NA 1468.

Stewart, Hilary,
1984 *Cedar: Tree of Life to the Northwest Coast Indians*, Douglas and McIntyre, Vancouver.
1977 *Indian Fishing: Early Methods on the Northwest Coast*, Douglas and McIntrye, Vancouver.

Suquamish Tribal Museum,
1989 *WATERBORNE: Gift of the Indian Canoe*, audio-visual production, Current-Rutledge, Seattle.

Suttles, Wayne (vol. editor),
1990 *Handbook of North American Indians*, Vol. 7, "Northwest Coast," Smithsonian Institution, Washington, D.C.
1987 *Coast Salish Essays*, University of Washington Press, Seattle.
1974 "Economic Life of the Coast Salish of Haro and Rosario Straits," in *American Indian Ethnohistory: Coast Salish Indians*, Garland Publishing, New York.

Swan, J.G.
1857 *The Northwest Coast, or Three Years' Residence in Washington Territory*, (Reprinted 1972) University of Washington Press, Seattle.

Vaughan, Thomas,
1971 *Paul Kane, The Columbian Wanderer Sketches, Paintings and Comments, 1846-1847*, Oregon Historical Society, Portland.

Waterman, T.T.
1930 "The Paraphernalia of the Duwamish Spirit Canoe Ceremony," in *Indian Notes*, Museum of the American Indian, Heye Foundation, New York, Vol. 7, No. 2-4.
1922 "Geographic Names" in *Geographic Review*, April.
1920 *The Whaling Equipment of the Makah Indians*, University of Washington Publications in Anthropology, 1: 1-67.

Waterman, T.T. and Geraldine Coffin,
1920 *Types of Canoes on Puget Sound*, Museum of the American Indian, Heye Foundation, New York.

DUGOUTS IN WASHINGTON

CAMP NORWESTER, Lopez Island
35' Haida canoe built by Bill Holm in 1968,
24' Kwakiutl type built by Bill Holm in 1958,
26' Two-person racing canoe MOUNTAIN BOY,
42' Six-person racing canoe MALAHAT ROSE.

CENTER FOR WOODEN BOATS, Seattle
32' Nootkan style built by S. Martin in 1987,
15'8" Coast Salish canoe,
9'6" Nootkan style child's canoe,
16' Nootkan style variation.

DUWAMISH TRIBE, Seattle
Approximately 20' shovel-nose built
by Frank Fowler in 1989.

FORT NISQUALLY POINT DEFIANCE PARK, Tacoma
Nootkan style canoe made in 1940s.

ISLAND COUNTY HISTORICAL SOCIETY, Coupeville
Four fully restored racing canoes,
50' eleven-person racing canoe TELEGRAPH, built in 1906 by the Edwards brothers.

HOH RESERVATION
25' Sealing canoe built by Charlie Hudson.

JEFFERSON COUNTY HISTORICAL SOCIETY, Port Townsend
20' Nootkan style canoe on display at Point Hudson.

LOWER ELWHA S'KLALLAM RESERVATION
32' Coast Salish craft NA-KE-TAH,
built by Richard Mike in 1989.

LUMMI RESERVATION
50' TELEGRAPH II racing canoe,
50' WINTER HAWK racing canoe,
Many other 50' eleven-person, 42' six-person,
26' two-person and one-person racing craft.

MAKAH CULTURE AND RESEARCH CENTER, Neah Bay
Four Nootkan style canoes (32', 25', 16' & 10')
built by L. Wilkie and S. Brown in 1975.
Numerous older racing and fishing dugouts.

MARYMOOR MUSEUM, Redmond
Two dugout canoes.

MUCKLESHOOT RESERVATION
20' Shovel-nose built by L. "Doc" Starr and Marvin Starr, Sr.

NISQUALLY RESERVATION
20' Nootkan style,
18' Shovel-nose canoe.

NOOKSACK RESERVATION
42' Six-person racing canoe,
built by G. Swanaset.

PIONEER FARM MUSEUM, Eatonville
Shovel-nose canoe built by Ev Woodward.

PORT GAMBLE S'KLALLAM RESERVATION
35' Adapted Coast Salish style built
by Duane Pasco and Jake Jones in 1989,
16' Sda'kwihl built by Pasco and Jones in 1988.

PORT ANGELES NATIONAL PARK VISITOR'S CENTER, Port Angeles
25' Nootkan hunting canoe.

QUINAULT RESERVATION
26' Motorized river racing canoes.

QUILEUTE RESERVATION & TRIBAL SCHOOL, La Push
54' Nootkan style canoe,
43' "Whaling" canoe and
24' "Cargo" canoe, each built by David Forlines.
Numerous older 25' motorized fishing and racing dugouts crafted by 'Ribs' Penn and relatives.

ROCKPORT STEELHEAD MUSEUM, Rockport
40' Shovel-nose canoe.

SAMISH TRIBAL PRESERVATION FUND, Anacortes
27' Salish canoe carved by T. Powell
and J. Waterhouse, Jr.

SKAGIT COUNTY HISTORICAL MUSEUM, La Conner
40' Shovel-nose dugout,
24' Shovel-nose canoe built by A. Joe in 1908, under cover along the Swinomish Canal.

SKOKOMISH RESERVATION
50' CHIEF JOHN eleven-person racing canoe,
42' SKOKWON six-person racing craft.

SNOQUALMIE VALLEY HISTORICAL MUSEUM, North Bend
Three Shovel-nose canoes on pedestals under eves.

STEILACOOM TRIBAL MUSEUM, Steilacoom
27' Nootkan style canoe equipped with an outboard motor.

SUQUAMISH RESERVATION & MUSEUM, Suquamish
50' Racing canoe built by Waterhouse and R. Edwards,
42' Six-person racing canoe,
25' Nootkan style sealing canoe built by
Quileute carver Laven Coe, on loan from Kitsap County Historical Society,
13' Coast Salish canoe on loan from Burke Museum.
Numerous one-& two-person racing dugouts.

SWINOMISH RESERVATION
50' Racing canoe.

THOMAS BURKE WASHINGTON STATE MEMORIAL MUSEUM, Seattle
Small Snohomish canoe acquired in 1941 by
Roger Ernesti,
37' Northern style canoe built before 1896,
13'8" Haida canoe built by J. Wallace,
35' Shovel-nose,
Small double-ender,
2 one-person "singles" racing canoes from Snoqualmie.

TULALIP RESERVATION, Marysville
50' KLA-HOW-YA eleven-person racing craft,
36' Nootkan canoe, once owned by Chief
William Shelton,
36' Nootkan style replica of Shelton's canoe, built by
Gerry Jones and Joe Gobins in 1989,
(Also in Tulalip: Steve Philipp's 20' dugout
& collection of carved models).

UPPER SKAGIT RESERVATION
50' ROLLING THUNDER eleven-person racing canoe,
42' ARMINA ROSE six-person racing canoe.

WASHINGTON STATE HISTORICAL SOCIETY, Tacoma
Nootkan and Northern style canoes.

NOTE: Many other small dugouts exist—particularly racing canoes on Tribal Reservations.